# Higher Symmetries and Its Application in Microwave Technology, Antennas and Metamaterials

# Higher Symmetries and Its Application in Microwave Technology, Antennas and Metamaterials

Special Issue Editors

**Guido Valerio**
**Oscar Quevedo-Teruel**

MDPI • Basel • Beijing • Wuhan • Barcelona • Belgrade

**MDPI**

*Special Issue Editors*

Guido Valerio
Sorbonne Université
France

Oscar Quevedo-Teruel
KTH Royal
Institute of Technology
Sweden

*Editorial Office*
MDPI
St. Alban-Anlage 66
4052 Basel, Switzerland

This is a reprint of articles from the Special Issue published online in the open access journal *Symmetry* (ISSN 2073-8994) in 2019 (available at: https://www.mdpi.com/journal/symmetry/special_issues/ Higher_Symmetries_Its_Application_Microwave_Technology_Antennas_Metamaterials).

For citation purposes, cite each article independently as indicated on the article page online and as indicated below:

LastName, A.A.; LastName, B.B.; LastName, C.C. Article Title. *Journal Name* **Year**, *Article Number*, Page Range.

**ISBN 978-3-03921-876-9 (Pbk)**
**ISBN 978-3-03921-877-6 (PDF)**

Cover image courtesy of Adrian Tamayo-Dominguez.

# Contents

# About the Special Issue Editors

**Guido Valerio** received his PhD degree in electromagnetics from the Sapienza University of Rome, Rome, Italy, in 2009. From 2011 to 2014, he was a Researcher with the IETR, Rennes, France. Since 2014, he has been an Associate Professor at Sorbonne Université, Paris, France. He was a Visiting Scholar at the University of Houston in 2008 and at the University of Michigan in 2015, 2016, and 2017.

His research interests include numerical methods for wave propagation and scattering in complex structures, leaky-wave antennas, SIW, and multi-layered media. He is currently working on wave propagation along artificial surfaces having geometrical higher symmetries. He is the co-author of more than 50 papers in international journals, and more than 100 in international conferences. He was the co-organizer of three convened sessions on higher symmetries (at AP-S 2017, EuCAP 2018, and META 2017). He is currently the Chair of the COST Action CA18223 on higher-symmetric artificial materials.

Dr. Valerio was a recipient of the Leopold B. Felsen Award in 2008. In 2010, he was a recipient of the Barzilai Prize for the best paper at the National Italian Congress of Electromagnetism (XVIII RiNEm). In 2014, he was a recipient of the Raj Mittra Travel Grant for junior researchers presented at the IEEE Antennas and Propagation Society Symposium, Memphis, TN, USA. In 2018, he was co-author of the best paper in "Electromagnetic and Antenna theory" at the 12th European Conference on Antennas and Propagation (EuCAP), London, UK.

**Oscar Quevedo-Teruel** received his degree in Telecommunication Engineering from Carlos III University of Madrid Spain in 2005, part of which was done at Chalmers University of Technology in Gothenburg, Sweden. He obtained his PhD from Carlos III University of Madrid in 2010 and was then invited as a postdoctoral researcher at the University of Delft (The Netherlands). In 2010–2011, Dr. Quevedo-Teruel joined the Department of Theoretical Physics of Condensed Matter at Universidad Autonoma de Madrid as a research fellow and went on to continue his postdoctoral research at Queen Mary University of London in 2011–2013.

In 2014, he joined the Electromagnetic Engineering Division, in the School of Electrical Engineering and Computer Science at KTH Royal Institute of Technology in Stockholm, Sweden, where he is an Associate Professor and director of the Master Programme in Electromagnetics Fusion and Space Engineering. He has been an Associate Editor of the IEEE Transactions on Antennas and Propagation since 2018, and he is the delegate of EurAAP for Sweden, Norway, and Iceland for the period 2018–2020. He is a distinguished lecturer of the IEEE Antennas and Propagation Society for the period 2019–2021.

He has made scientific contributions to higher symmetries, transformation optics, lens antennas, metasurfaces, leaky-wave antennas, multimode microstrip patch antennas, and high impedance surfaces. He is the co-author of 77 papers in international journals, 139 at international conferences, and has received approval on 3 patents.

# Preface to "Higher Symmetries and Its Application in Microwave Technology, Antennas and Metamaterials"

Artificial materials composed of periodic arrangement of inclusions (unit cells) in a host medium have been widely studied and used in photonics and microwaves in the last few decades. Recently, it has been remarked that unit cells exhibiting specific symmetries can lead to dispersive properties never found before: ultra large bandwidth of operation, significantly reduced losses due to dielectric-less propagation, miniaturization effects, and enhanced stopband for electromagnetic bandgaps. These properties can meet the expectation of new communication devices providing beam-scanning antennas, high isolation, miniaturization, high-power capabilities, and reconfigurability.

When the unit cell of a periodic structure is invariant under a geometrical transformation, the structure is said to be "higher symmetric", since the higher symmetry is the minimal symmetry defining the entire structure starting from a part of the unit cell. Higher-symmetric structures along one direction were first introduced in the 1970s, and many fundamental results on the subject were summarized in the seminal paper (Hessel et al., *Proc. IEEE*, 1973). Namely, glide symmetry (a composition of a mirroring and a translation) and twist symmetry (a composition of a rotation and a translation) were analyzed there with reference to periodic metallic waveguides. Those results did not spark new research until today, when recent studies on artificial materials have provided fields of application not considered before. It has been remarked that glide symmetry can enable negative-index dispersion when otherwise not possible (Quesada et al., *Opt. Lett.*, 2014) and a higher effective refractive index for lens applications in (T. Chang et al., *Nature Comm.*, 2016) and (Cavallo and Felita, *IEEE Trans. Antennas Propag.*, 2017). 2-D glide-symmetric metasurfaces can considerably reduce the dispersion of wave propagation with respect to simple periodic structures, mimicking an artificial material whose refractive index remains almost invariant over a wideband (O. Quevedo-Teruel et al., *IEEE Antennas Wireless Propag. Lett.*, 2016) and (O. Quevedo-Teruel et al., *IEEE Antennas Wireless Propag. Lett.*, 2018). Furthermore, 2-D glide-symmetric metasurfaces improve both the bandwidth and the attenuation of stopbands with respect to simple periodic structures, offering new solutions for integrated circuits and gap waveguide technology (M. Ebrahimpouri et al., *IEEE Trans. Microw. Theory Techn.*, 2018). The need for modeling and explaining the dispersive effects related to these symmetries has led to numerical and analytical solutions based on mode-matching (Valerio et al., *IEEE Trans. Microw. Theory Techn.*, 2018) and equivalent circuits (Valerio et al., *IEEE Antennas Propag.*, 2017). The effects of symmetries on the inter-cell coupling have been quantified in (Bagheriasl et al., *IEEE Trans. Microw. Theory Techn.*, 2019).

This Special Issue considers the role of symmetries of spatial or temporal nature in different kinds of electromagnetic problems. The first paper "Time-reversal Symmetry in Antenna Theory" discusses how time-reversal invariance affects properties of lossless radiating systems. Namely, antenna matching conditions are examined, and for the first time, the generation of the time-reversed field distribution of a radiating system is proposed, by illuminating a matched antenna with a suitable Herglotz wave. The following four papers deal with symmetric configurations for the design of radiating and guided-wave devices. In the paper "Fully Metallic Flat Lens Based on Locally Twist-Symmetric Array of Complementary Split-Ring Resonators", a twisted arrangement of split-ring resonators is used to synthetize different effective refractive index by modifying the order of the twist symmetry. This leads to a flat lens collimating a spherical wave into a plane wave. In "Analysis of Periodic Structures Made of Pins Inside a Parallel Plate Waveguide", wave propagation inside a structure made of periodic metallic pins is studied. The pins position is modified starting from a fully-symmetric configuration to a glide-symmetric one, and by tuning in a symmetric or asymmetric way the pin length. The effects on the frequency-dispersion and on the stop-band frequency width are studied with full-wave simulations. The paper "One-Plane Glide-Symmetric Holey Structures for Stop-Band and Refraction Index Reconfiguration" proposes a holey metasurface whose slots are of elliptical shape and exhibit in-plane glide symmetry, in order to minimize misalignment problems in off-plane in glide-symmetric structures.

In "Twist and Glide Symmetries for Helix Antenna Design and Miniaturization", twist and glide symmetries are used in order to miniaturize a helix antenna. A combination of glide and twist symmetry is also defined by including glide corrugations along a helicoidal pattern.

The last two papers of the Special Issue present methods to solve an electromagnetic problem in a domain defined by higher-symmetric boundary conditions. In "Modeling of Glide-Symmetric Dielectric Structures", a mode matching is proposed to study glide-symmetric structures made with dielectric inclusions. The field distribution of different modes is computed in both fully-symmetric and glide-symmetric configurations. Finally, in "Bloch Analysis of Electromagnetic Waves in Twist-Symmetric Lines", a multimodal characterization of a suitable sub-unit cell is used in order to recover the dispersion diagram of the twist-symmetric line, by highlighting the role of inter-cell coupling due to higher-order modes.

**Guido Valerio, Oscar Quevedo-Teruel**
*Special Issue Editors*

![symmetry logo] *symmetry*

MDPI

*Article*

# Time-reversal Symmetry in Antenna Theory

**Mário G. Silveirinha**

Instituto Superior Técnico and Instituto de Telecomunicações, University of Lisbon, Avenida Rovisco Pais, 1, 1049-001 Lisboa, Portugal; mario.silveirinha@co.it.pt

Received: 24 February 2019; Accepted: 2 April 2019; Published: 4 April 2019

**Abstract:** Here, I discuss some implications of the time-reversal invariance of lossless radiating systems. I highlight that time-reversal symmetry provides a rather intuitive explanation for the conditions of polarization and impedance matching of a receiving antenna. Furthermore, I describe a solution to generate the time-reversed electromagnetic field through the illumination of a matched receiving antenna with a Herglotz wave.

**Keywords:** Time-reversal symmetry; Antennas; Lorentz reciprocity

## 1. Introduction

Time reversal is the operation that flips the arrow of time such that $t \to -t$ [1,2]. Remarkably, the laws that rule the microscopic dynamics of most physical systems are invariant under a time-reversal transformation (the exceptions occur in some nuclear interactions and are irrelevant in the context of this study). This property implies that under suitable initial conditions, the time reversed dynamics may be generated and observed in a real physical setting, similar to a movie played backwards. Ultimately, the invariance under time reversal implies that at a microscopic level the physical phenomena are intrinsically reversible: if a particular time evolution is compatible with the physical laws, then the time-reversed dynamics also is.

A consequence of "time-reversal invariance" is that the propagation of light in standard waveguides is inherently bi-directional, even if the system does not have any particular spatial symmetry. For example, if an electromagnetic wave can go through, a metallic pipe with no back-reflections, then the time-reversed wave also can, but propagating in the opposite direction. This rather remarkable property is usually explained with the help of the Lorentz reciprocity theorem [3,4], but it is ultimately a consequence of microscopic reversibility and time reversal invariance [2,5,6].

In this article, I reexamine the consequences of time-reversal invariance in antenna theory. I show that time-reversal invariance provides a rather intuitive explanation for the conditions of impedance and polarization matching in the theory of the receiving antenna. In addition, I prove that in a time-harmonic regime the time-reversed wave can be generated through the illumination of the receiving antenna with a superposition of plane waves generated in the far-field.

## 2. Time-Reversal Symmetry

It is well known that the equations of *macroscopic* electrodynamics are not time reversal invariant when the system has dissipative elements. This is so because the description provided by macroscopic electrodynamics is incomplete, as it only models the time evolution of the electromagnetic degrees of freedom. In contrast, in a microscopic formalism –with all the light and matter degrees of freedom included in the analysis– the system dynamics is time-reversal symmetric. Thus, in some sense, macroscopic dissipative systems (e.g., lossy dielectrics) have a *hidden* time-reversal symmetry [6]. To circumvent this complication, here I will focus on systems with negligible material absorption, so that the dynamics determined by the macroscopic Maxwell equations are time-reversal symmetric.

## 2.1. General Case

Consider the propagation of light in some lossless, dispersion-free, dielectric system described by the Maxwell equations:

$$\nabla \times \mathbf{E} = -\mu_0 \frac{\partial \mathbf{H}}{\partial t},$$
$$\nabla \times \mathbf{H} = \mathbf{j} + \varepsilon \frac{\partial \mathbf{E}}{\partial t} \tag{1}$$

with $\varepsilon = \varepsilon(\mathbf{r})$. The time-reversal operation $\mathcal{T}$ transforms the electromagnetic fields $\mathbf{E}, \mathbf{H}$ and the current density $\mathbf{j}$ as $\mathbf{E} \xrightarrow{\mathcal{T}} \mathbf{E}^{TR}$, $\mathbf{H} \xrightarrow{\mathcal{T}} \mathbf{H}^{TR}$, and $\mathbf{j} \xrightarrow{\mathcal{T}} \mathbf{j}^{TR}$ with [2]:

$$\mathbf{E}^{TR}(\mathbf{r}, t) = \mathbf{E}(\mathbf{r}, -t),$$
$$\mathbf{H}^{TR}(\mathbf{r}, t) = -\mathbf{H}(\mathbf{r}, -t),$$
$$\mathbf{j}^{TR}(\mathbf{r}, t) = -\mathbf{j}(\mathbf{r}, -t). \tag{2}$$

The transformed fields satisfy the same equations as the original fields. Under a time-reversal transformation the magnetic field and the current density flip sign, whereas the electric field does not. Thus, the former are said to be odd under a time-reversal transformation, whereas the latter is even. As a consequence, the Poynting vector $\mathbf{S} = \mathbf{E} \times \mathbf{H}$ also flips sign under a time reversal operation, so that the wave dynamics and direction of propagation are effectively reversed. The time reversal symmetry is rather general and applies to waves with an *arbitrary* variation in time.

For example, consider the scenario illustrated in Figure 1a, which represents a scattering problem with two waveguides connected by some arbitrary junction (two-port microwave network). The two incoming waves $\mathbf{E}_1^+$ and $\mathbf{E}_2^+$ can have arbitrary time variations and their scattering originates two outgoing waves, $\mathbf{E}_1^-$ and $\mathbf{E}_2^-$. As illustrated in Figure 1b, the time-reversal operation swaps the roles of the incoming and outgoing waves, because it flips the direction of propagation. Hence, the time-reversed signals are given by $\mathbf{E}_i^{TR,\pm}(\mathbf{r}, t) = \mathbf{E}_i^{\mp}(\mathbf{r}, -t)$. In particular, suppose that some wave incident in port 1 is fully transmitted to port 2. Then, if port 2 is illuminated with the time-reversed transmitted signal it will reproduce the original signal in port 1, but reversed in time. Thereby, time-reversal invariant systems are intrinsically bi-directional, independent of any spatial asymmetry.

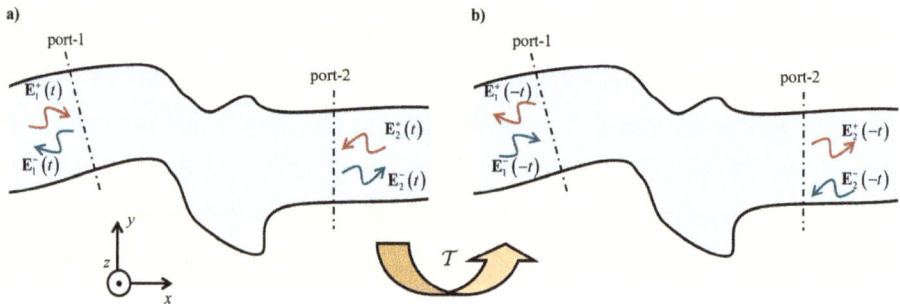

**Figure 1.** Illustration of the effect of the time-reversal operation in a time-domain scattering problem: (a) The incoming waves $\mathbf{E}_1^+$ and $\mathbf{E}_2^+$ are scattered by the junction and originate two outgoing waves $\mathbf{E}_1^-$ and $\mathbf{E}_2^-$. (b) Time-reversed scenario where the roles of the incoming and outgoing waves are exchanged.

The enunciated results can be generalized in a straightforward way to dispersive lossless dielectrics, e.g., to material structures characterized by some real-valued scalar permittivity $\varepsilon = \varepsilon(\omega, \mathbf{r})$ (e.g., a Lorentz dispersive model with no dissipation). The reason is that the electrodynamics of lossless dispersive systems can be formulated as a Schrödinger-type time evolution problem [7–9], which in the case of reciprocal media (e.g., standard dielectrics) is time-reversal invariant.

Furthermore, as discussed in Reference [10], most *lossless* nonlinear systems are time-reversal symmetric and hence are also bi-directional (see also Reference [11] for the acoustic case). For example, for an instantaneous Kerr-type nonlinear response with $\varepsilon = \varepsilon_0(\chi^{(1)} + \chi^{(3)}\mathbf{E}\cdot\mathbf{E})$, the Maxwell equations (1) remain time-reversal invariant. Interestingly, the time-harmonic response of a two-port microwave network with nonlinear components is generically asymmetric [12–14]. Indeed, if the ports are individually excited by the *same* time-harmonic signal the level of the transmitted signal depends on which port is excited; in this sense, nonlinear systems are nonreciprocal as the transmissivity is generically direction dependent for a given time-harmonic excitation [12–14]. In summary, lossless nonlinear systems are usually both time-reversal invariant and nonreciprocal, the two conditions are not incompatible [10]. In the rest of the article, I focus on linear systems.

### 2.2. Time-Harmonic Variation

Consider a time-harmonic solution of the Maxwell equations, such that the electromagnetic fields and current density are of the form: $\mathbf{E}(\mathbf{r}, t) = \mathrm{Re}\{\mathbf{E}_\omega(\mathbf{r})e^{j\omega t}\}$, $\mathbf{H}(\mathbf{r}, t) = \mathrm{Re}\{\mathbf{H}_\omega(\mathbf{r})e^{j\omega t}\}$, $\mathbf{j}(\mathbf{r}, t) = \mathrm{Re}\{\mathbf{j}_\omega(\mathbf{r})e^{j\omega t}\}$, with $\omega$ being the real-valued oscillation frequency. Under a time reversal the electric field is transformed as $\mathbf{E}(\mathbf{r}, t) \to \mathrm{Re}\{\mathbf{E}_\omega(\mathbf{r})e^{-j\omega t}\} = \mathrm{Re}\{\mathbf{E}_\omega^*(\mathbf{r})e^{j\omega t}\}$, where the symbol "*" stands for complex conjugation. Hence, the complex amplitudes of the fields and current density are transformed as:

$$\begin{aligned} \mathbf{E}_\omega(\mathbf{r}) &\xrightarrow{\mathcal{T}} \mathbf{E}_\omega^*(\mathbf{r}), \\ \mathbf{H}_\omega(\mathbf{r}) &\xrightarrow{\mathcal{T}} -\mathbf{H}_\omega^*(\mathbf{r}), \\ \mathbf{j}_\omega(\mathbf{r}) &\xrightarrow{\mathcal{T}} -\mathbf{j}_\omega^*(\mathbf{r}). \end{aligned} \tag{3}$$

Thus, in the frequency domain the time-reversal operation is closely linked to phase conjugation [15,16]. Similarly, voltages and currents are transformed as:

$$\begin{aligned} V_\omega &\xrightarrow{\mathcal{T}} V_\omega^*, \\ I_\omega &\xrightarrow{\mathcal{T}} -I_\omega^*. \end{aligned} \tag{4}$$

For example, consider a $N$-port microwave network such that the voltages and currents at a generic port $i$ are of the form: $V_{\omega,i} = V_{\omega,i}^+ + V_{\omega,i}^-$ and $I_{\omega,i} = (V_{\omega,i}^+ - V_{\omega,i}^-)/Z_0$, $i = 1, \dots, N$. Here, $V_{\omega,i}^+$ represents an incoming (incident) wave and $V_{\omega,i}^-$ an outgoing (scattered) wave. The characteristic impedance of the ports is $Z_0$. The incident and scattered waves are related as $\mathbf{V}^- = \mathbf{S}\cdot\mathbf{V}^+$, where $\mathbf{V}^\pm = \left[V_{\omega,i}^\pm\right]$ are column vectors and $\mathbf{S} = [S_{ij}]$ is the scattering matrix. The time reversal operation exchanges the roles of the incident and scattered waves, such that $\mathbf{V}^{\mathrm{TR},\pm} = \mathbf{V}^{\mp,*}$. Therefore, if the system is time-reversal invariant $\mathbf{V}^{+,*} = \mathbf{S}\cdot\mathbf{V}^{-,*}$. Thus, the scattering matrix must satisfy $\mathbf{S} = \mathbf{S}^{-1,*}$. On the other hand, for a lossless system the incident power must equal the scattered power: $\mathbf{V}^-\cdot\mathbf{V}^{-,*} = \mathbf{V}^+\cdot\mathbf{V}^{+,*}$. To satisfy this additional constraint the scattering matrix must be unitary $\mathbf{S}\cdot\mathbf{S}^\dagger = 1$. Combining the two results, one finds that the scattering matrix must be transpose symmetric:

$$\mathbf{S} = \mathbf{S}^T. \tag{5}$$

Thus, any time-reversal invariant linear lossless system is necessarily reciprocal ($S_{ij} = S_{ji}$) [17].

Here, I note in passing that in electromagnetic theory the time-reversal operator $\mathcal{T}$ is idempotent, such that $\mathcal{T}^2 = 1$. In other words, a "double" time reversal leaves the system dynamics unchanged. In contrast, in condensed matter theory the time reversal operator satisfies $\mathcal{T}^2 = -1$, and because of this property the scattering matrix of fermionic systems is anti-symmetric, $\mathbf{S} = -\mathbf{S}^T$ [17]. It was recently shown that photonic systems protected by a special parity-time-duality ($\mathcal{PTD}$) symmetry are constrained by $\mathbf{S} = -\mathbf{S}^T$, and thereby are matched at all ports ($S_{ii} = 0$) [17]. Such systems can enable bi-directional transmission of light free of back scattering.

## 3. Application to Antenna Theory

The time-reversal property may be used to explain several well-known properties of radiating systems. Similar to the previous section, I assume that the antennas are formed by lossless materials, e.g., lossless dielectrics or perfect conductors. In particular, the radiation efficiency of the antennas is 100%.

Consider a generic antenna radiating in free-space (Figure 2a). The antenna is fed by a generator with a time-harmonic variation. The antenna radiates the electromagnetic fields $\mathbf{E}_\omega^{\text{rad}}, \mathbf{H}_\omega^{\text{rad}}$. By definition, the antenna impedance is $Z_a = V_{0,\omega}/I_{0,\omega}$ where $V_{0,\omega}, I_{0,\omega}$ are the complex amplitudes of the voltage and current at the antenna terminals. In the far-field region the radiated electric field is asymptotically of the form [18]:

$$\mathbf{E}_\omega^{\text{rad}} \approx \mathbf{E}_\omega^{\text{ff}} \equiv \eta_0 j k_0 I_{0,\omega} \frac{e^{-jk_0 r}}{4\pi r} \mathbf{h}_e(\hat{\mathbf{r}})$$
$$\mathbf{h}_e(\hat{\mathbf{r}}) = \hat{\mathbf{r}} \times (\hat{\mathbf{r}} \times \frac{1}{I_{0,\omega}} \int \mathbf{j}_\omega(\mathbf{r}') e^{jk_0 \hat{\mathbf{r}} \cdot \mathbf{r}'} d^3 \mathbf{r}').$$

(6)

**a)**                                        **b)**

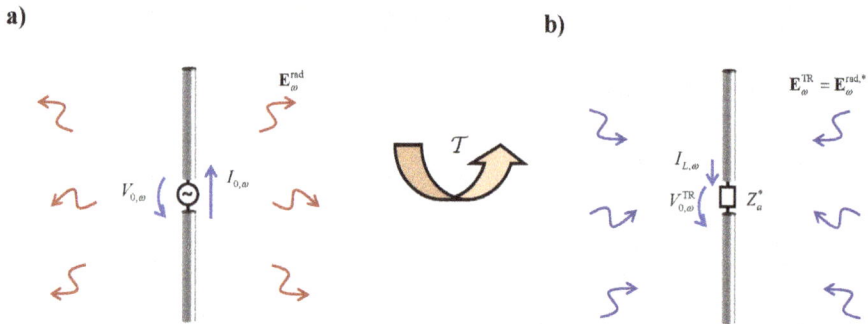

**Figure 2.** (**a**) An antenna fed by a time-harmonic generator radiates in free-space. (**b**) Time-reversed problem wherein all the radiated energy returns to the antenna. The antenna terminals are connected to a matched load.

In the above, $k_0 = \omega/c$ is the free-space wave number, $\eta_0$ is the free-space impedance, and $\mathbf{h}_e(\hat{\mathbf{r}})$ is the (vector) effective height of the antenna, which depends on the direction of observation $\hat{\mathbf{r}}$ ($\hat{\mathbf{r}}$ can be expressed in terms of angles $\theta$, $\varphi$ associated with a system of spherical coordinates). The antenna effective height depends on the total current distribution $\mathbf{j}_\omega(\mathbf{r}')$, which includes the external currents associated with the generator and the induced polarization and conduction currents in the materials.

The polarization of the antenna in the direction $\hat{\mathbf{r}}$ is determined by the closed curve defined by $\mathbf{E}^{\text{rad}}(t) = \text{Re}\{\mathbf{E}_\omega^{\text{rad}} e^{j\omega t}\} \sim \text{Re}\{\mathbf{h}_e(\hat{\mathbf{r}}) e^{j\omega t}\}$, and hence by the effective height $\mathbf{h}_e$ because the electric field is evaluated in the far-field region.

### 3.1. Polarization and Impedance Matching

Consider now the time-reversed problem represented in Figure 2b, where all the radiated energy is returned back to the antenna. The time reversed voltage and current at the antenna terminals are $V_{0,\omega}^{\text{TR}} = V_{0,\omega}^*$ and $I_{0,\omega}^{\text{TR}} = -I_{0,\omega}^*$ (Equation (4)). The current flowing into the antenna terminals (inward direction) is $I_{L,\omega} = -I_{0,\omega}^{\text{TR}}$ (see Figure 2; note that in the scenario of Figure 2a the current is positive when it flows in the outward direction). From here, it follows that $V_{0,\omega}^{\text{TR}}/I_{L,\omega} = V_{0,\omega}^*/I_{0,\omega}^* = Z_a^*$, i.e., in the time-reversed scenario the generator is effectively equivalent to a matched load with impedance $Z_a^*$. Furthermore, the field arriving to the antenna in the direction $\hat{\mathbf{r}}$ is evidently $\mathbf{E}_\omega^{\text{TR}} = \mathbf{E}_\omega^{\text{rad},*} \sim \mathbf{h}_e^*(\hat{\mathbf{r}})$, which is the well-know condition for polarization matching. These properties show that in the time-reversed problem the antenna is *impedance matched* to the load and *polarization matched* to the incident wave for any direction $\hat{\mathbf{r}}$.

Thus, the time-reversal invariance provides a rather intuitive understanding of the conditions of impedance and polarization matching, as it shows that the two conditions emerge naturally in the time-reversed problem where the receiving antenna captures the energy arriving from the far-field with 100% efficiency.

In the time domain $\mathbf{E}^{\text{TR}}(t) = \mathbf{E}^{\text{rad}}(-t)$ and thereby the polarization curve associated with $\mathbf{E}_\omega^{\text{TR}}$ is the same as the polarization curve associated with $\mathbf{E}_\omega^{\text{rad}}$. In a time period $(T = 2\pi/\omega)$, $\mathbf{E}^{\text{TR}}(t), \mathbf{E}^{\text{rad}}(t)$ follow the same polarization curve but in *opposite* directions due to the time-reversal link. Yet, the polarization of the two waves is the same, i.e., the antenna and the wave are polarization matched, because the propagation directions of the two waves differ by a minus sign ($\mathbf{E}_\omega^{\text{rad}}$ propagates in the outward radial direction and $\mathbf{E}_\omega^{\text{TR}}$ in the inward radial direction).

For example, an antenna that radiates a right-circularly polarized (RCP) wave in some direction of space is polarization matched to an incoming plane wave with RCP polarization. Even though the wave and antenna polarizations are identical, the geometrical senses of rotation of the relevant electric fields are *opposite*. This otherwise intriguing property can be understood as a simple consequence of time-reversal invariance.

*3.2. Time-Reversed Field Generated with a Far-Field Illumination*

The problem of generating a time-reversed field distribution is of practical interest, as it enables concentrating and focusing energy from the far field into some desired region of space. The theory and application of time-reversed fields were developed and extensively explored by Fink and co-authors [19–24]. Here, I revisit the problem and highlight some features that were not discussed in Reference [24].

In the time-reversed problem of Figure 2b the incident wave $\mathbf{E}_\omega^{\text{TR}}$ propagates from $r = \infty$ to the antenna where it is fully absorbed by the matched load, without generating any back-reflections. It is natural to wonder what happens if the same antenna is illuminated by the time-reversed far field (time reversal of $\mathbf{E}_\omega^{\text{ff}}$) rather than by the fully time-reversed field (the time reversal of $\mathbf{E}_\omega^{\text{rad}}$ given by $\mathbf{E}_\omega^{\text{TR}}$). In the former case, the incident wave $\mathbf{E}_\omega^{\text{inc}}(\mathbf{r})$ should be a superposition of plane waves emerging from all possible directions of space $\hat{\mathbf{r}}'$. From Equation (6), the field $d\mathbf{E}_\omega^{\text{inc}}(\mathbf{r})$ associated with the wave emerging from the infinitesimal solid angle $d\Omega(\hat{\mathbf{r}}')$ must have amplitude proportional to $e^{+jk_0\hat{\mathbf{r}}'\cdot\mathbf{r}}\mathbf{h}_e^*(\hat{\mathbf{r}}')d\Omega(\hat{\mathbf{r}}')$.

Notably, I prove in the Appendix A that the solution of the scattering problem formulated in the previous paragraph can be constructed from the fully time reversed field $\mathbf{E}_\omega^{\text{TR}}$. Specifically, when an *impedance-matched* antenna is illuminated by the incident field

$$\mathbf{E}_\omega^{\text{inc}}(\mathbf{r}) = \frac{k_0^2}{8\pi^2} I_{0,\omega}^* \eta_0 \int e^{+jk_0\hat{\mathbf{r}}'\cdot\mathbf{r}} \mathbf{h}_e^*(\hat{\mathbf{r}}')d\Omega(\hat{\mathbf{r}}'), \tag{7}$$

the field scattered by the antenna is precisely given by $\mathbf{E}_\omega^{\text{scat}} = \mathbf{E}_\omega^{\text{TR}} - \mathbf{E}_\omega^{\text{inc}}$, such that the total field is $\mathbf{E}_\omega^{\text{TR}}$. Thus, $\mathbf{E}_\omega^{\text{TR}}$ may be both understood as an incident wave that is absorbed by the antenna with no back-scattering, or alternatively as the superposition of an incident wave ($\mathbf{E}_\omega^{\text{inc}}$) and the corresponding field back scattered by the antenna ($\mathbf{E}_\omega^{\text{scat}}$). The two cases, even though totally different from a physical point of view, cannot be mathematically distinguished in time-harmonic regime.

As previously mentioned, the incident field $\mathbf{E}_\omega^{\text{inc}}$ is a superposition of propagating plane waves emerging from all directions of space. This type of wave is known as a Herglotz wave. The integral in Equation (7) is over all solid angles $d\Omega(\hat{\mathbf{r}}')$. Furthermore, it is shown in Appendix A that the scattered field has the following asymptotic form in the far-field region:

$$\mathbf{E}_\omega^{\text{scat}}(\mathbf{r}) \approx -\eta_0 jk_0 I_{0,\omega}^* \frac{e^{-jk_0 r}}{4\pi r} \mathbf{h}_e^*(-\hat{\mathbf{r}}). \tag{8}$$

Comparing Equations (6) and (8) it is evident that the power scattered by the antenna when it is illuminated by $\mathbf{E}_\omega^{\text{inc}}$ is $P_{\text{scat}} = P_{\text{rad}}$ where $P_{\text{rad}}$ is the power radiated by the antenna in the scenario of Figure 2a. Furthermore, since the total field $\mathbf{E}_\omega^{\text{inc}} + \mathbf{E}_\omega^{\text{scat}}$ is identical to $\mathbf{E}_\omega^{\text{TR}}$ it is evident that the power absorbed by the matched load is $P_r = P_{\text{rad}}$. These properties imply that when the impedance-matched antenna is illuminated by the Herglotz wave it captures the same power as it scatters: $P_r = P_{\text{scat}}$. The property $P_r = P_{\text{scat}}$ is specific to the Herglotz wave considered here, and it generally does not hold true for other far-field excitations [25,26]. Note that the polarization curve associated with the scattered field $\mathbf{E}_\omega^{\text{scat}}$ along the direction $\hat{\mathbf{r}}$ is determined by $\mathbf{h}_e^*(-\hat{\mathbf{r}})$, which generally differs from the polarization of the antenna in transmitting mode.

It is emphasized that the fully time reversed field $\mathbf{E}_\omega^{\text{TR}}$ can be excited simply by illuminating the impedance matched antenna with the Herglotz wave $\mathbf{E}_\omega^{\text{inc}}$, which is a superposition of propagating plane waves.

## 4. Conclusions

I revisited the topic of time-reversal symmetry in macroscopic electromagnetism. I showed that under a time-reversal transformation a transmitting antenna becomes the impedance matched receiving antenna. Heuristically, the excitation with the time-reversed wave must be the most effective way of delivering power to an antenna. Thus, the time-reversal invariance provides a simple and intuitive understanding of the conditions of impedance and polarization matching in antenna theory. In particular, it elucidates why a polarization matched incident wave has an electric field that rotates geometrically in a direction opposite to that of the field radiated by the antenna in the same direction. In addition, I generalized the ideas of Reference [24] and showed that the time reversal of the field emitted by a lossless transmitting antenna can be created by illuminating an impedance matched receiving antenna with the far-field excitation associated with the Herglotz wave given by the Equation (7). In such a scenario, the power captured by the matched load is precisely the same as the power scattered by the antenna.

**Funding:** This research was funded by the Institution of Engineering and Technology (IET) under the A F Harvey Engineering Research Prize and by Fundação para Ciência e a Tecnologia (FCT) under project PTDC/EEITEL/4543/2014 and UID/EEA/50008/2019.

**Conflicts of Interest:** The authors declare no conflict of interest.

## Appendix A

Consider the configuration of Figure 2a, where a generic lossless antenna radiates in free-space. Let $\mathbf{j}_\omega(\mathbf{r}')$ be the total electric current distribution, determined both by the external current associated with the generator and by the polarization and conduction currents in the materials. The radiated fields in time-harmonic regime may be expressed in terms of a vector potential as

$$\mathbf{A}_\omega(\mathbf{r}) = \mu_0 \int \mathbf{j}_\omega(\mathbf{r}') \frac{e^{-jk_0|\mathbf{r}-\mathbf{r}'|}}{4\pi|\mathbf{r}-\mathbf{r}'|} d^3\mathbf{r}'. \tag{A1}$$

Under a time reversal, the vector potential is transformed as $\mathbf{A}_\omega(\mathbf{r}) \xrightarrow{\mathcal{T}} -\mathbf{A}_\omega^*(\mathbf{r})$. Thus, the time-reversed vector potential is:

$$\mathbf{A}_\omega^{\text{TR}}(\mathbf{r}) = \mu_0 \int -\mathbf{j}_\omega^*(\mathbf{r}') \frac{e^{+jk_0|\mathbf{r}-\mathbf{r}'|}}{4\pi|\mathbf{r}-\mathbf{r}'|} d^3\mathbf{r}'. \tag{A2}$$

Using $e^{+jk_0|\mathbf{r}-\mathbf{r}'|} = e^{-jk_0|\mathbf{r}-\mathbf{r}'|} + 2j\sin(k_0|\mathbf{r}-\mathbf{r}'|)$, I obtain the decomposition $\mathbf{A}_\omega^{\text{TR}} = \mathbf{A}_\omega^{\text{inc}} + \mathbf{A}_\omega^{\text{scat}}$, with

$$\mathbf{A}_\omega^{\text{scat}}(\mathbf{r}) = \mu_0 \int -\mathbf{j}_\omega^*(\mathbf{r}') \frac{e^{-jk_0|\mathbf{r}-\mathbf{r}'|}}{4\pi|\mathbf{r}-\mathbf{r}'|} d^3\mathbf{r}', \tag{A3a}$$

$$\mathbf{A}_\omega^{\text{inc}}(\mathbf{r}) = \mu_0 \int -\mathbf{j}_\omega^*(\mathbf{r}') \frac{j\sin(k_0|\mathbf{r}-\mathbf{r}'|)}{2\pi|\mathbf{r}-\mathbf{r}'|} d^3\mathbf{r}'. \tag{A3b}$$

Evidently, the time-reversed field has a similar decomposition $\mathbf{E}_\omega^{\text{TR}} = \mathbf{E}_\omega^{\text{inc}} + \mathbf{E}_\omega^{\text{scat}}$ (see also Reference [24]). The field $\mathbf{E}_\omega^{\text{scat}}$ is obtained from $\mathbf{A}_\omega^{\text{scat}}$, and thus satisfies the Sommerfeld radiation conditions. Thus, $\mathbf{E}_\omega^{\text{scat}}$ can be understood as the wave scattered by $\mathbf{E}_\omega^{\text{inc}}$. From Equation (A3a) it is simple to check that in the far-field region

$$\mathbf{E}_\omega^{\text{scat}}(\mathbf{r}) \approx -jk_0\eta_0 \frac{e^{-jk_0 r}}{4\pi r} \hat{\mathbf{r}} \times \left(\hat{\mathbf{r}} \times \int \mathbf{j}_\omega^*(\mathbf{r}')e^{+jk_0\hat{\mathbf{r}}\cdot\mathbf{r}'}d^3\mathbf{r}'\right). \tag{A4}$$

Comparing this result with Equation (6), one obtains Equation (8).

The potential $\mathbf{A}_\omega^{\text{inc}}$ is an analytic function and can be written as a superposition of plane waves. Indeed, from

$$\frac{\sin k_0 r}{4\pi r} = \frac{k_0}{16\pi^2} \int e^{-jk_0\hat{\mathbf{k}}\cdot\mathbf{r}}d\Omega(\hat{\mathbf{k}}), \tag{A5}$$

the incident vector potential may be expressed as:

$$\mathbf{A}_\omega^{\text{inc}}(\mathbf{r}) = -\mu_0 \frac{jk_0}{8\pi^2} \int d\Omega(\hat{\mathbf{k}})e^{jk_0\hat{\mathbf{k}}\cdot\mathbf{r}} \left(\int d^3\mathbf{r}' \, \mathbf{j}_\omega^*(\mathbf{r}')e^{-jk_0\hat{\mathbf{k}}\cdot\mathbf{r}'}\right). \tag{A6}$$

With the help of Equation (6), it can be checked that the "incident" electric field $\mathbf{E}_\omega^{\text{inc}} = (1/j\omega\varepsilon_0)\nabla \times \nabla \times \mathbf{A}_\omega^{\text{inc}}/\mu_0$ is given by Equation (7).

The power received by an impedance-matched antenna is

$$P_{\text{r}} = \frac{|V_{\text{oc}}|^2}{8R_a}. \tag{A7}$$

Here, $R_a = \text{Re}\{Z_a\}$ is the input resistance of the antenna and $V_{\text{oc}}$ is the voltage induced by the incident field at the antenna terminals when they are terminated with an open circuit. As is well known, for reciprocal systems the open-circuit voltage is $V_{\text{oc}} = \mathbf{E}_0^{\text{inc}} \cdot \mathbf{h}_e(\hat{\mathbf{r}})$ where $\mathbf{E}_0^{\text{inc}}$ is field associated with an incident plane wave (arriving from direction $\hat{\mathbf{r}}$) evaluated at the origin [18]. Thus, from the superposition principle, the voltage induced by the Herglotz wave given by Equation (7) is:

$$V_{\text{oc}} = \frac{k_0^2}{8\pi^2} I_{0,\omega}^* \eta_0 \int |\mathbf{h}_e(\hat{\mathbf{r}})|^2 d\Omega(\hat{\mathbf{r}}). \tag{A8}$$

For a lossless system the input resistance is coincident with the radiation resistance, which from (6) can be written as $R_a = \eta_0 \frac{k_0^2}{16\pi^2}\int |\mathbf{h}_e(\hat{\mathbf{r}})|^2 d\Omega(\hat{\mathbf{r}})$. This result implies that $V_{\text{oc}} = 2I_{0,\omega}^* R_a$ so that the received power is given by $P_{\text{r}} = \frac{1}{2}R_a|I_{0,\omega}|^2 = P_{\text{rad}}$. This direct analysis confirms that when the antenna is illuminated by the Herglotz wave the power absorbed by a matched load is exactly the power radiated by the antenna in the scenario of Figure 2a.

## References

1. Feynman, R.; Leighton, R.; Sand, M. *The Feynman Lectures on Physics*; California Institute of Technology: Pasadena, CA, USA, 1963.
2. Casimir, H.B.G. Reciprocity theorems and irreversible processes. *Proc. IEEE* **1963**, *51*, 1570. [CrossRef]
3. Potton, R.J. Reciprocity in optics. *Rep. Prog. Phys.* **2004**, *67*, 717–754. [CrossRef]
4. Caloz, C.; Alù, A.; Tretyakov, S.; Sounas, D.; Achouri, K.; Deck-Léger, Z.L. Electromagnetic Nonreciprocity. *Phys. Rev. Appl.* **2018**, *10*, 047001. [CrossRef]
5. Onsager, L. Reciprocal relations in irreversible processes I. *Phys. Rev.* **1931**, *37*, 405.
6. Silveirinha, M.G. Hidden Time-Reversal Symmetry in Dissipative Reciprocal Systems. *Opt. Express* **2019**, in press. Available online: https://arxiv.org/abs/1903.02944 (accessed on 3 April 2019).

7.  Silveirinha, M.G. Chern Invariants for Continuous Media. *Phys. Rev. B* **2015**, *92*, 125153. [CrossRef]
8.  Silveirinha, M.G. Topological classification of Chern-type insulators by means of the photonic Green function. *Phys. Rev. B* **2018**, *97*, 115146. [CrossRef]
9.  Silveirinha, M.G. Modal expansions in dispersive material systems with application to quantum optics and topological photonics. 2018. Available online: https://arxiv.org/abs/1712.04272 (accessed on 3 April 2019).
10. Fernandes, D.E.; Silveirinha, M.G. Asymmetric Transmission and Isolation in Nonlinear Devices: Why They Are Different. *IEEE Antennas Wirel. Propag. Lett.* **2018**, *17*, 1953. [CrossRef]
11. Tanter, M.; Thomas, J.L.; Coulouvrat, F.; Fink, M. Breaking of time reversal invariance in nonlinear acoustics. *Phys. Rev. E* **2001**, *64*, 016602. [CrossRef]
12. Shadrivov, I.V.; Fedotov, V.A.; Powell, D.A.; Kivshar, Y.S.; Zheludev, N.I. Electromagnetic wave analogue of an electronic diode. *New J. Phys.* **2011**, *13*, 033025.
13. Mahmoud, A.M.; Davoyan, A.R.; Engheta, N. All-passive nonreciprocal metastructure. *Nat. Commun.* **2015**, *6*, 8359. [CrossRef]
14. Sounas, D.L.; Soric, J.; Alù, A. Broadband Passive Isolators Based on Coupled Nonlinear Resonances. *Nat. Electron.* **2018**, *1*, 113–119. [CrossRef]
15. Maslovski, S.; Tretyakov, S. Phase conjugation and perfect lensing. *J. Appl. Phys.* **2003**, *94*, 4241. [CrossRef]
16. Pendry, J.B. Time Reversal and Negative Refraction. *Science* **2008**, *322*, 71–73. [CrossRef] [PubMed]
17. Silveirinha, M.G. PTD symmetry protected scattering anomaly in optics. *Phys. Rev. B* **2017**, *95*, 035153. [CrossRef]
18. Balanis, C.A. *Antenna Theory*, 4th ed.; John Wiley & Sons, Inc.: Hoboken, NJ, USA, 2016.
19. Cassereau, D.; Fink, M. Time-Reversal of Ultrasonic Fields-Part III: Theory of the Closed Time-Reversal Cavity. *IEEE Trans. Ultrason. Ferroelectr. Freq.* **1992**, *39*, 579–592. [CrossRef] [PubMed]
20. Fink, M. Time-Reversed Acoustics. *Sci. Am.* **1999**, *281*, 91–97. [CrossRef]
21. de Rosny, J.; Fink, M. Overcoming the Diffraction Limit in Wave Physics Using a Time-Reversal Mirror and a Novel Acoustic Sink. *Phys. Rev. Lett.* **2002**, *89*, 124301. [CrossRef] [PubMed]
22. Lerosey, G.; de Rosny, J.; Tourin, A.; Derode, A.; Montaldo, G.; Fink, M. Time Reversal of Electromagnetic Waves. *Phys. Rev. Lett.* **2004**, *92*, 193904. [CrossRef]
23. Lerosey, G.; de Rosny, J.; Tourin, A.; Fink, M. Focusing Beyond the Diffraction Limit with Far-Field Time Reversal. *Science* **2007**, *315*, 1120. [CrossRef] [PubMed]
24. de Rosny, J.; Lerosey, G.; Fink, M. Theory of Electromagnetic Time-Reversal Mirrors. *IEEE Trans. Antennas Propag.* **2010**, *58*, 3139–3149. [CrossRef]
25. Andersen, J.B.; Vaughan, R.G. Transmitting, receiving and scattering properties of antennas. *IEEE Antennas Propag. Mag.* **2003**, *45*, 93–98. [CrossRef]
26. Andersen, J.B.; Frandsen, A. Absorption Efficiency of Receiving Antennas. *IEEE Trans. Antennas Propag.* **2005**, *53*, 2843–2849. [CrossRef]

*symmetry*

MDPI

*Article*

# Fully Metallic Flat Lens Based on Locally Twist-Symmetric Array of Complementary Split-Ring Resonators

Oskar Dahlberg [1,*], Guido Valerio [2] and Oscar Quevedo-Teruel [1]

[1] Division of Electromagnetic Engineering, KTH Royal Institute of Technology, 100 44 Stockholm, Sweden; oscarqt@kth.se

[2] Laboratoire d'Électronique et Électromagnétisme, Sorbonne Université, F-75005 Paris, France; guido.valerio@sorbonne-universite.fr

* Correspondence: oskdah@kth.se

Received: 14 March 2019; Accepted: 19 April 2019; Published: 21 April 2019

**Abstract:** In this article, we demonstrate how twist symmetries can be employed in the design of flat lenses. A lens design is proposed, consisting of 13 perforated metallic sheets separated by an air gap. The perforation in the metal is a two-dimensional array of complementary split-ring resonators. In this specific design, the twist symmetry is local, as it is only applied to the unit cell of the array. Moreover, the twist symmetry is an approximation, as it is only applied to part of the unit cell. First, we demonstrate that, by varying the order of twist symmetry, the phase delay experienced by a wave propagating through the array can be accurately controlled. Secondly, a lens is designed by tailoring the unit cells throughout the aperture of the lens in order to obtain the desired phase delay. Simulation and measurement results demonstrate that the lens successfully transforms a spherical wave emanating from the focal point into a plane wave at the opposite side of the lens. The demonstrated concepts find application in future wireless communication networks where fully-metallic directive antennas are desired.

**Keywords:** twist symmetry; lens antenna; complementary split-ring resonator; complementary split ring resonator (CSRR)

## 1. Introduction

A periodic structure possesses a higher geometrical symmetry if it is invariant under a translation and one or more additional geometrical operations. One-dimensional (1D) structures possessing higher symmetries were studied during the 1960s and 1970s [1–4]. In those early studies, two types of higher symmetries were investigated, namely Cartesian glide and twist (also called screw) symmetry. A Cartesian glide symmetry is obtained if the additional geometrical operation, which is applied jointly with the translation, is a mirroring with respect to a plane. More specifically, a structure possesses Cartesian glide symmetry if its unit cell consists of two sub-unit cells that are displaced a distance $p/2$ and mirrored with respect to a plane, where $p$ is the period of the full unit cell. A twist symmetry is obtained if the additional geometrical operation is a rotation around the periodicity axis. A structure possesses an $m$-fold twist symmetry ($m$ being an integer) if its unit cell consists of $m$ sub-unit cells, which are displaced a distance $p/m$ and rotated $2\pi/m$ with respect to the adjacent sub-unit cells. A two-fold twist-symmetric structure also possesses glide symmetry if the sub-unit cell is mirror-symmetric with respect to one plane that includes the periodicity axis. Moreover, a purely-periodic structure, i.e., a structure without a higher symmetry, possesses one-fold twist symmetry ($m = 1$). In [1–4], it was shown that there were no stop-bands between the $m$ first modes in the dispersion diagram ($m = 2$ in the case of glide). This modal characteristic has been applied for the design of 1D glide-symmetric forward and backward scanning leaky-wave antennas [5–7].

Recently, higher symmetries, and glide symmetry in particular, have received a renewed attention in the microwave community, as higher symmetric structures provide attractive properties for the design of microwave components [8–19]. These properties are especially attractive in the mm-wave regime, where dielectric losses are prohibitive, and higher symmetries permit a control of the propagation characteristics on fully-metallic periodic structures [20]. More specifically, it has been demonstrated that higher symmetric structures increase the equivalent refractive index [21–23] and reduce the dispersion [9,10,16,17] of conventional periodic structures. These properties have recently been applied in the design of two-dimensional (2D) fully-metallic glide-symmetric wideband metasurface lenses [9,10]. Moreover, glide-symmetric structures have demonstrated that, while they suppress the stop-band between the first two modes, a huge stop-band is present between the second and third modes [12]. This property has been used to reduce the leakage between waveguide interconnections [13] and design cost-efficient gap waveguides at mm-wave frequencies [14,15]. In order to reduce the computation time and give valuable physical insight, several methods for calculating the dispersion diagram of glide-symmetric structures have been presented [24–28].

Although similar properties have been demonstrated in twist-symmetric structures [16–18,24], very few twist-symmetric devices have been conceived. In twist-symmetric structures, in contrast to glide-symmetric structures, the *m*-fold order of symmetry presents an additional degree of freedom in the design. Twist symmetry has been successfully employed in the design of reconfigurable filters [17], compact phase shifters [19], miniaturization of helix antennas [29], and polarization transformers [30]. In this work, we demonstrate how the additional control of the propagation characteristics in twist-symmetric structures can be applied in the design of flat lenses.

The paper is organized as follows: In Section 2, two simulation studies are conducted. First, in Section 2.1, local twist symmetry is investigated, and the effect of adding different orders of twist symmetries in the structure is highlighted. After, in Section 2.2, local twist symmetries are employed in the design of a flat lens. In Section 3, the results are discussed. Finally, in Section 4, the methods employed in the study and the procedure undertaken are explained.

## 2. Results

Our study is divided in two steps. First, in Section 2.1, a general study of twist-symmetric complementary split ring resonators (CSRRs) is conducted. In this initial study, the effects of locally adding an approximate twist symmetry to an array of CSRRs is highlighted. Secondly, in Section 2.2, a lens design consisting of an array of tailored CSRRs is conducted based on the results obtained in Section 2.1.

### 2.1. Study of Twist-Symmetric Complementary Split Ring Resonators

In this work, four different structures with local twist symmetry are studied. The structures are illustrated in Figure 1. In this initial study, the structures are 3D periodic. The different unit cells consisted of different numbers of layers of perforated metallic sheets of a thickness $t$, separated by an air gap of thickness $h$. The perforation in the metal consisted of two concentric semi-circular slots, also called CSRRs, of radius, $R$. The slots had a width $sw$, and a metallic bridge of width $g$ connected the metal on each side of the slot. The studied structures were 1-, 3-, 4-, and 6-fold twist-symmetric with 1, 3, 4, and 6 metallic sheets per unit cell, respectively. The twist symmetry was local since only the unit cell was twist-symmetric, and not the full array of CSRRs. Furthermore, the twist symmetry was approximate since only the slots were rotated between adjacent sub-unit cells, and not the entire sub-unit cell. Moreover, since the sub-unit cell exhibited mirror symmetry, the rotation between two adjacent sub-unit cells in the smallest geometrical full unit cell was halved. This means that the rotation between two adjacent sub-unit cells was not $2\pi/m$, but rather $\pi/m$, where $m$ is the order of the twist symmetry. Consequently, the rotation between adjacent sub-unit cells was 180°, 60°, 45°, and 30° in the 1-, 3-, 4-, and 6-fold twist-symmetric structures, respectively.

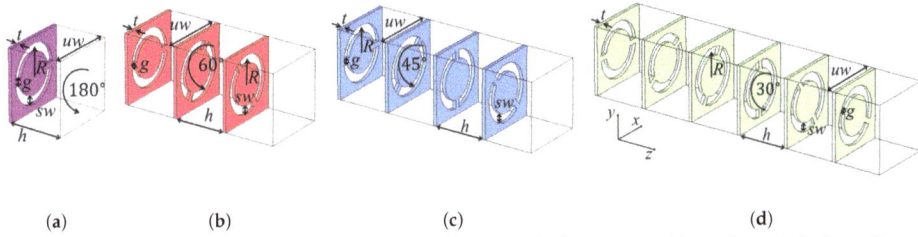

(a)        (b)        (c)        (d)

**Figure 1.** Simulated CSRRs with local twist symmetry. Studied structures: (**a**) purely periodic ($m = 1$), (**b**) 3-fold, (**c**) 4-fold, and (**d**) 6-fold.

In Figure 2a, the propagation constants for the four structures are represented. In all structures, the lateral periodicity, $uw$, was 20 mm, the slots were placed at a radius $R = 7$ mm; the slot width, $sw$, was 3 mm, and the metallic bridge separating the slots had a width of $g = 1.5$ mm. The separation between the metallic sheets, $h$, was 4 mm, and the thickness of the metallic sheets, $t$, was 1 mm. The final implementation (i.e., the lens) will operate with the second mode of the CSRR array. Therefore, only this mode is plotted in Figure 2a. The electric field profile of the second mode is illustrated in the inset of Figure 2a. The mode was odd with respect to the center of the unit cell, in contrast to the first mode, which was even. The odd field distribution was consistent with the excitation employed in the final implementation of the lens. Notably, the electric field pattern in the slot resembled that of a $TE_{10}$-mode in a rectangular waveguide. In fact, the dispersion characteristics of the mode were similar to the ones in a periodically-loaded rectangular waveguide.

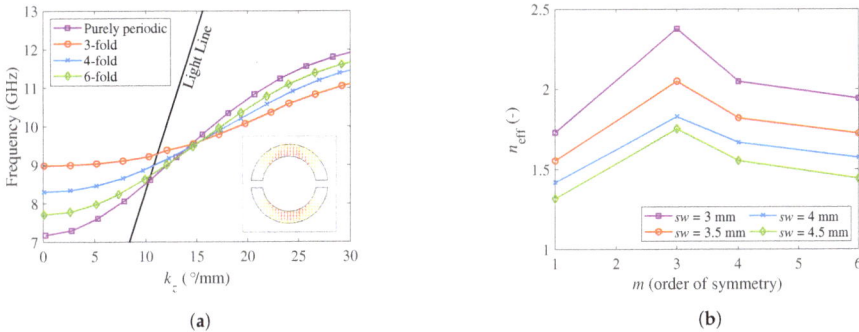

(a)        (b)

**Figure 2.** (**a**) Simulated propagation constant with dimensions: $uw = 20$ mm, $R = 7$ mm, $sw = 3$ mm, $g = 1.5$ mm, $t = 1$ mm, and $h = 4$ mm. (**b**) Simulated effective refractive index at 11 GHz with dimensions: $uw = 20$ mm, $R = 7$ mm, $g = 1.5$ mm, $t = 1$ mm, and $h = 4$ mm.

From the results of Figure 2a, we can conclude that the phase delay can be controlled by changing the order of the symmetry. In previous works, it has been demonstrated that by increasing the order of twist symmetry, an increased density can be obtained [16–18,24]. However, in Figure 2a, the highest density is not necessarily achieved in the structure with the highest order of the symmetry. The reason for this discrepancy with previously-reported results is that, in this study, the sub-unit cell period was kept constant, in contrast to previous studies where the full unit cell periodicity was kept constant. Moreover, the lowest order of twist symmetry (i.e., the three-fold structure) was the most different from the purely-periodic structure. There were two reasons for this. First, the cut-off frequency of the CSRR was shifted upwards in the twist-symmetric structures compared to the purely-periodic structure. The CSRRs in different layers of the structures effectively formed a waveguide with a cut-off frequency. When adjacent layers were rotated, the effective width of the waveguide was reduced, leading to an up-shift in the cut-off frequency. The largest rotation between the two sub-unit cells was obtained in the three-fold structure, and hence, the shift was more severe in this configuration, as the effective width of the waveguide was the smallest. Secondly, the slope of the dispersion curve was more gradual in the twist-symmetric structures compared to the purely-periodic structure. The more gradual slope was caused by the reduced coupling between the CSRRs in different layers for the twist-symmetric structures, compared to the purely-periodic structure. The reduced coupling resulted in a narrower pass band for the second mode [30].

In Figure 2b, the effective refractive index at 11 GHz is presented. The order of symmetry and slot width were swept. The remaining parameters were: $R = 7.5$ mm, $uw = 20$ mm, $g = 2$ mm, $h = 4$ mm, and $t = 1$ mm. Again, the density of a periodic structure can be controlled by varying the order of the symmetry. The increased density was related to the increased path traveled by the mode, which was forced to revolve in a helical manner around the periodicity axis. The smallest pitch of the helix, and consequently the highest density, was obtained in the three-fold structure. In fact, for increasing order of twist symmetry, we approached the purely-periodic structure as the rotation between subsequent sub-unit cells was decreasing. Additionally, if another geometrical parameter was allowed to vary simultaneously (the slot width, $sw$, in this case), a large continuous range of effective indices can be obtained. This control of the phase delay has been previously used to design compact phase shifters [19]. Here, we employed this effect for the design of a flat lens.

## 2.2. Lens Design Using CSRRs

The operation of the lens was conceptually similar to the one of transmit arrays [31–34]. The phase delay experienced by a wave propagating through the lens was controlled throughout the aperture. The full lens structure is presented in Figure 3. The lens consisted of 13 perforated metallic sheets with a thickness of 1 mm. The sheets were made of aluminum and were separated by 4 mm of air. These sheets were thick enough to remain flat in a practical realization of the lens. However, due to the manufacturing process employed here (laser cutting), the sheets deformed significantly. Therefore, a layer a 4 mm-thick Rohacell 51HF ($\varepsilon_r = 1.065$) was added for structural support in between the metallic layers in the realized prototype. The manufactured lens (in the measurement setup) is illustrated in Figure 3c.

The perforation in the metallic sheets consisted of an array of CSRRs. The CSRRs were tailored throughout the aperture so the lens provided the required phase correction necessary to transform a spherical wave emanating from the focal point into a plane wave at the opposite side. A similar configuration has been employed to obtain a wideband linear-to-linear polarization transformation with very low insertion losses [30]. However, in that work, a normally incident plane wave was assumed, and the CSRRs remained unchanged throughout the aperture. Moreover, the first and last layers were different in order to produce a polarization transformation.

Thirteen metallic layers were employed here to produce 3-, 4-, and 6-fold twist-symmetric unit cells. In this way, each twist configuration can fit an integer number of periods into 12 metallic layers. One additional layer, identical to the first, was inserted at the end to ensure that the lens was symmetric and no polarization transformation was performed. If desired, polarization transformation can be integrated into the structure if the first and last layers are different. The phase delay throughout the lens was tailored so that a spherical wave emanating from the focal point, $fp$, arrived in phase at the opposite side of the lens. To obtain this, $\psi(r) = l(r) + T \cdot n_{\text{eff}}(r)$ must remain constant throughout the aperture (up to an integer addition of free-space wavelengths), where $r$ is the radial coordinate in the aperture, $l(r) = fp/\cos[\tan^{-1}(r/fp)]$ is the total optical path from the focal point through the lens, and $n_{\text{eff}}(r)$ is the effective refractive index of the lens at the position $r$. The focal point in the designed lens was 130 mm, and the width of the lens, $w$, was 220 mm. The total thickness of the lens, $T$, was 61 mm.

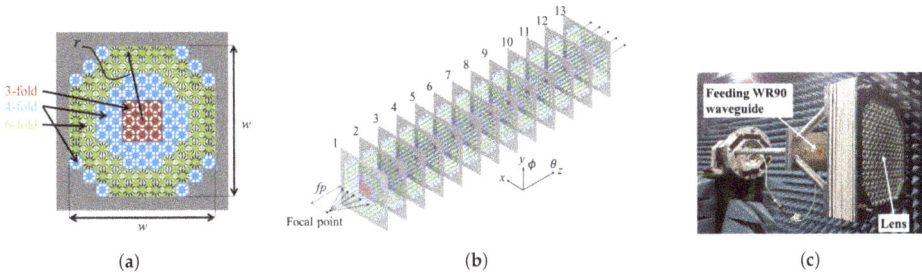

**Figure 3.** (a–b) Layout of the lens: (a) front view, and (b) perspective view. The distance between the metallic sheets is exaggerated for increased clarity. (c) Measurement setup in the anechoic chamber.

The simulated electric field for the lens, at 11 GHz, excited by a half-wavelength dipole, is presented in Figure 4a,b for the E-plane and H-plane. The lens successfully transformed the spherical wave into a plane wave. To estimate how successful the transformation was, the normalized radiation pattern at 11 GHz, both for the lens fed with a rectangular waveguide (WR90) and for the isolated waveguide (normalized to the maximum realized gain of the full lens antenna), is plotted in Figure 4c,d for the E-plane and H-plane. A clear improvement in the gain of roughly 6 dB was achieved with the lens. The measured H-plane radiation pattern is included in Figure 4d, and the measurement corroborated the simulation. The measured E-plane radiation pattern was distorted by the struts necessary for mounting the antenna in the anechoic chamber. Therefore, the E-plane cut is not presented here. The losses in the metallic sheets were below 1% of the stimulated power in simulations.

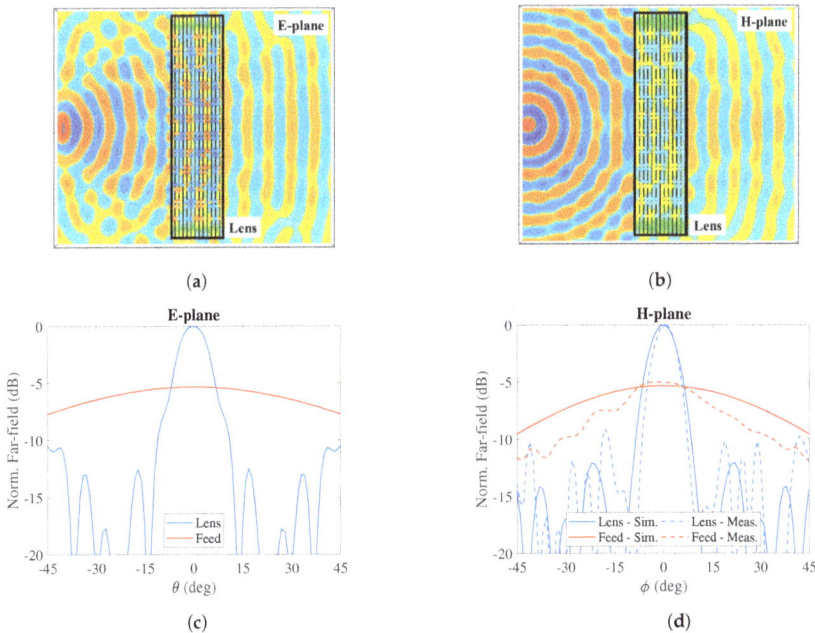

**Figure 4.** Simulated E-field and far-field of the lens. (a–b) x-component of the electric field excited by an x-oriented half-wavelength dipole placed in the focal point of the lens: (a) E-plane and (b) H-plane. (c–d) Normalized radiation pattern of the lens excited with a rectangular waveguide (WR90) placed at the focal point of the lens: (c) E-plane and (d) H-plane. The measurement of the H-plane cut is included as well. The width, $w$, of the lens is 220 mm, and the focal point is 130 mm from the first layer. The separation between the sheets is 4 mm, and the thickness of the metallic sheets is 1 mm.

*Symmetry* **2019**, *11*, 581

## 3. Discussion

In this work, a periodic structure with local twist symmetry was applied for the design of a fully-metallic lens. First, it was demonstrated that, by introducing different orders of an approximated twist symmetry into the unit cell of a 2D array of CSRRs, the phase delay can be controlled. Moreover, an excellent range of different phase delays can be achieved by only varying the symmetry order. Secondly, a lens was designed by varying the order of the symmetry throughout the aperture, obtaining the required phase delay at each point. The lens consisted of 13 perforated metallic sheets separated by an air gap. The perforation in each sheet was an array of CSRRs. The CSRRs were rotated in each consecutive layer to obtain the twist symmetry. The first and last layers of the lens were identical in order to ensures that no polarization transformation was performed in the lens. However, a polarization transformer can be integrated into the structure by using different orientation of the CSRRs in the first and last layer. Full-wave simulation results showed that the lens structure successfully transformed a spherical wave emanating from the focal point of the lens into a plane wave at the opposite side. The presented concepts can be employed in future design intended for wireless communication networks at high frequencies where fully-metallic directive antennas are desired [20].

## 4. Materials and Methods

CST Microwave Studio [35] has been used to produce the simulation results in this article. In the study conducted in Section 2.1, the structures were simulated in the Eigenmode Solver of CST, with periodic boundary conditions in all directions. In the study conducted in Section 2.2, the structure was simulated in the Time-Domain Solver of *CST*. The measurements were carried out in an anechoic chamber. The lens was fed with a WR90 standard waveguide.

**Author Contributions:** Conceptualization, O.Q.-T.; formal analysis, O.D. and G.V.; methodology, O.D.; visualization, O.D.; writing, original draft preparation, O.D.; writing, review and editing, O.D., O.Q.-T. and G.V.; supervision, O.Q.-T. and G.V.

**Funding:** This research was partly funded by The French National Research Agency Grant Number ANR-16-CE24-0030, partly by the Vinnova project High-5 (2018-01522), under the Strategic Programme on Smart Electronic Systems, and partly by the Stiftelsen Åforsk project H-Materials (18-302).

**Conflicts of Interest:** The authors declare no conflict of interest.

## References

1. Hessel, A.; Chen, M.H.; Li, R.C.M.; Oliner, A.A. Propagation in Periodically Loaded Waveguides with Higher Symmetries. *Proc. IEEE* **1973**, *61*, 183–195. [CrossRef]
2. Crepeau, P.J.; McIsaac, P.R. Consequences of Symmetry in Periodic Structures. *Proc. IEEE* **1963**, *52*, 33–43. [CrossRef]
3. Mittra, R.; Laxpati, S. Propagation in a Wave Guide with Glide Reflection Symmetry. *Can. J. Phys.* **1965**, *43*, 353–372. [CrossRef]
4. Kieburtz, R.; Impagliazzo, J. Multimode Propagation on Radiating Traveling-Wave Structures with Glide-Symmetric Excitation. *IEEE Trans. Antennas Propag.* **1970**, *18*, 3–7. [CrossRef]
5. Cao, W.; Chen, Z.N.; Hong, W.; Zhang, B.; Liu, A. A Beam Scanning Leaky-Wave Slot Antenna with Enhanced Scanning Angle Range and Flat Gain Characteristic Using Composite Phase-Shifting Transmission Line. *IEEE Trans. Antennas Propag.* **2014**, *62*, 5871–5875. [CrossRef]
6. Wu, J.J.; Wu, C.J.; Hou, D.J.; Liu, K.; Yang, T.J. Propagation of Low-Frequency Spoof Surface Plasmon Polaritons in a Bilateral Cross-Metal Diaphragm Channel Waveguide in The Absence of Bandgap. *IEEE Photonics J.* **2015**, *7*, 1–8. [CrossRef]
7. Lyu, Y.L.; Liu, X.X.; Wang, P.Y.; Erni, D.; Wu, Q.; Wang, C.; Kim, N.Y.; Meng, F.Y. Leaky-Wave Antennas Based on Noncutoff Substrate Integrated Waveguide Supporting Beam Scanning From Backward To Forward. *IEEE Trans. Antennas Propag.* **2016**, *64*, 2155–2164. [CrossRef]
8. Quesada, R.; Martín-Cano, D.; García-Vidal, F.; Bravo-Abad, J. Deep-Subwavelength Negative-Index Waveguiding Enabled By Coupled Conformal Surface Plasmons. *Opt. Lett.* **2014**, *39*, 2990–2993. [CrossRef]

9.  Quevedo-Teruel, O.; Ebrahimpouri, M.; Kehn, M.N.M. Ultrawideband Metasurface Lenses Based on Off-Shifted Opposite Layers. *IEEE Antennas Wirel. Propag. Lett.* **2016**, *15*, 484–487. [CrossRef]

10. Quevedo-Teruel, O.; Miao, J.; Mattsson, M.; Algaba-Brazalez, A.; Johansson, M.; Manholm, L. Glide-Symmetric Fully Metallic Luneburg Lens for 5G Communications at $K_a$-Band. *IEEE Antennas Wirel. Propag. Lett.* **2018**, *17*, 1588–1592. [CrossRef]

11. Padilla, P.; Herrán, L.; Tamayo-Domínguez, A.; Valenzuela-Valdés, J.; Quevedo-Teruel, O. Glide Symmetry to Prevent The Lowest Stopband of Printed Corrugated Transmission Lines. *IEEE Microw. Wirel. Compon. Lett.* **2018**, *28*, 1–3. [CrossRef]

12. Ebrahimpouri, M.; Quevedo-Teruel, O.; Rajo-Iglesias, E. Design Guidelines for Gap Waveguide Technology Based on Glide-Symmetric Holey Structures. *IEEE Microw. Wirel. Compon. Lett.* **2017**, *27*, 542–544. [CrossRef]

13. Ebrahimpouri, M.; Brazalez, A.A.; Manholm, L.; Quevedo-Teruel, O. Using Glide-Symmetric Holes to Reduce Leakage Between Waveguide Flanges. *IEEE Microw. Wirel. Compon. Lett.* **2018**, *28*, 473–475. [CrossRef]

14. Ebrahimpouri, M.; Rajo-Iglesias, E.; Sipus, Z.; Quevedo-Teruel, O. Cost-Effective Gap Waveguide Technology Based on Glide-Symmetric Holey EBG Structures. *IEEE Trans. Microw. Theory Tech.* **2018**, *66*, 927–934. [CrossRef]

15. Rajo-Iglesias, E.; Ebrahimpouri, M.; Quevedo-Teruel, O. Wideband Phase Shifter in Groove Gap Waveguide Technology Implemented With Glide-Symmetric Holey EBG. *IEEE Microw. Wirel. Compon. Lett.* **2018**, *28*, 476–478. [CrossRef]

16. Dahlberg, O.; Mitchell-Thomas, R.; Quevedo-Teruel, O. Reducing the Dispersion of Periodic Structures with Twist and Polar Glide Symmetries. *Sci. Rep.* **2017**, *7*, 10136. [CrossRef]

17. Ghasemifard, F.; Norgren, M.; Quevedo-Teruel, O. Twist and Polar Glide Symmetries: An Additional Degree of Freedom to Control The Propagation Characteristics of Periodic Structures. *Sci. Rep.* **2018**, *8*, 11266. [CrossRef]

18. Dahlberg, O.; Ghasemifard, F.; Valerio, G.; Quevedo-Teruel, O. Propagation Characteristics of Periodic Structures Possessing Twist and Polar Glide Symmetries. *EPJ Appl. Metamater.* **2019**, in pressing. [CrossRef]

19. Quevedo-Teruel, O.; Dahlberg, O.; Valerio, G. Propagation in Waveguides With Transversal Twist-Symmetric Holey Metallic Plates. *IEEE Microw. Wirel. Compon. Lett.* **2018**, *28*, 1–3. [CrossRef]

20. Quevedo-Teruel, O.; Ebrahimpouri, M.; Ghasemifard, F. Lens Antennas for 5G Communications Systems. *IEEE Commun. Mag.* **2018**, *56*, 36–41. [CrossRef]

21. Cavallo, D.; Felita, C. Analytical Formulas for Artificial Dielectrics with Nonaligned Layers. *IEEE Trans. Antennas Propag.* **2017**, *65*, 5303–5311. [CrossRef]

22. Cavallo, D. Dissipation Losses in Artificial Dielectric Layers. *(Early Access) IEEE Trans. Antennas Propag.* **2018**, *66*. [CrossRef]

23. Chang, T.; Kim, J.U.; Kang, S.K.; Kim, H.; Kim, D.K.; Lee, Y.H.; Shin, J. Broadband Giant-Refractive-Index Material Based on Mesoscopic Space-Filling Curves. *Nat. Commun.* **2016**, *7*, 12661. [CrossRef]

24. Chen, Q.; Ghasemifard, F.; Valerio, G.; Quevedo-Teruel, O. Modeling and Dispersion Analysis of Coaxial Lines With Higher Symmetries. *IEEE Trans. Microw. Theory Tech.* **2018**, *66*, 1–8. [CrossRef]

25. Valerio, G.; Sipus, Z.; Grbic, A.; Quevedo-Teruel, O. Accurate Equivalent-Circuit Descriptions of Thin Glide-Symmetric Corrugated Metasurfaces. *IEEE Trans. Antennas Propag.* **2017**, *65*, 2695–2700. [CrossRef]

26. Valerio, G.; Ghasemifard, F.; Sipus, Z.; Quevedo-Teruel, O. Glide-Symmetric All-Metal Holey Metasurfaces for Low-Dispersive Artificial Materials: Modeling and Properties. *IEEE Trans. Microw. Theory Tech.* **2018**, *66*, 1–14. [CrossRef]

27. Ghasemifard, F.; Norgren, M.; Quevedo-Teruel, O. Dispersion Analysis of 2-D Glide-Symmetric Corrugated Metasurfaces Using Mode-Matching Technique. *IEEE Microw. Wirel. Compon. Lett.* **2018**, *28*, 1–3. [CrossRef]

28. Mesa, F.; Rodríguez-Berral, R.; Medina, F. On the Computation of the Dispersion Diagram of Symmetric One-Dimensionally Periodic Structures. *Symmetry* **2018**, *10*, 307. [CrossRef]

29. Palomares-Caballero, A.; Padilla, P.; Valenzuela-Valdes, J.; Quevedo-Teruel, O. Twist and Glide Symmetries for Helix Antenna Design and Miniaturization. *Symmetry* **2019**, *11*, 349. [CrossRef]

30. Wei, Z.; Cao, Y.; Fan, Y.; Yu, X.; Li, H. Broadband Polarization Transformation via Enhanced Asymmetric Transmission Through Arrays of Twisted Complementary Split-Ring Resonators. *Appl. Phys. Lett.* **2011**, *99*, 221907. [CrossRef]

*Symmetry* **2019**, *11*, 581

31. Pfeiffer, C.; Grbic, A. Millimeter-Wave Transmitarrays for Wavefront and Polarization Control. *IEEE Trans. Microw. Theory Tech.* **2013**, *61*, 4407–4417. [CrossRef]

32. Pfeiffer, C.; Grbic, A. Bianisotropic Metasurfaces for Optimal Polarization Control: Analysis and Synthesis. *Phys. Rev. Appl.* **2014**, *2*, 044011. [CrossRef]

33. Padilla, P.; Muñoz-Acevedo, A.; Sierra-Castañer, M. Passive Planar Transmit-Array Microstrip Lens for Microwave Purpose. *Microw. Opt. Technol. Lett.* **2010**, *52*, 940–947. [CrossRef]

34. Yeap, S.B.; Qing, X.; Chen, Z.N. 77-GHz Dual-Layer Transmit-Array for Automotive Radar Applications. *IEEE Trans. Antennas Propag.* **2015**, *63*, 2833–2837. [CrossRef]

35. CST Microwave Studio, Version: 2017. Available online: http://www.cst.com/ (accessed on 20 February 2019)

*symmetry*

MDPI

*Article*

# Analysis of Periodic Structures Made of Pins Inside a Parallel Plate Waveguide

Nafsika Memeletzoglou [1], Carlos Sanchez-Cabello [1], Francisco Pizarro-Torres [2] and Eva Rajo-Iglesias [1,*]

[1]   Department of Signal Theory and Communications, University Carlos III of Madrid, 28911 Leganés, Spain;
      nafsika.memeletzouglou@uc3m.es (N.M.); carlos.sanchez@uc3m.es (C.S.-C.)
[2]   Escuela de Ingeniería Eléctrica, Pontificia Universidad Católica de Valparaíso, Valparaíso 2362804, Chile;
      francisco.pizarro.t@pucv.cl
*    Correspondence: eva.rajo@uc3m.es; Tel.: +34-916248774

Received: 15 March 2019; Accepted: 17 April 2019; Published: 22 April 2019

**Abstract:** In this work, we have analyzed different versions of periodic structures made with metallic pins located inside a parallel plate waveguide (PPWG), varying the symmetry and disposition of the pins. The analysis focuses on two main parameters related to wave propagation. On one hand, we have studied how the different proposed structures can create a stopband so that the parallel plate modes can be used in gap waveguide technology or filtering structures. On the other hand, we have analyzed the dispersion and equivalent refractive index of the first propagating transverse electromagnetic mode (TEM). The results show how the use of complex structures made with pins in the top and bottom plates of a PPWG have no advantages in terms of the achieved stopband size. However, for the case of the propagating mode, it is possible to find less dispersive modes and a higher range of equivalent refractive indices when using double-pin structures compared to a reference case with single pins.

**Keywords:** bed of nails; glide symmetry; gap waveguide technology; dispersion; stopband

## 1. Introduction

Periodic structures made with metallic pins have been used by the electromagnetism community in different kinds of applications. The original structure known as "the bed of nails" [1] can be considered a metasurface and is able to provide extraordinary boundary conditions. This property was used in the definition of a new technology known as a gap waveguide [2,3], where the structure is combined with a top metal lid to provide the cutoff frequency or stopband for all the parallel plate modes [4]. In this case, the pin structure acts as an artificial magnetic conductor (AMC) within a given frequency range, producing a stopband. This same concept was also used as a successful packaging solution [5] using not only pins but also other versions of symmetrical and periodic "pin-like" structures [6–8]. A similar idea was also implemented using dielectric materials [9] rather than metal.

Gap waveguide technology is currently a reality and its use in designing antennas and components has been widely extended [10]. Alternatives to pin structures to create the stopband in a PPWG have emerged, such as half-size pins [11] and holes [12,13], mainly with the purpose of providing easier manufacturing of this technology.

On the other hand, these pin structures have been also used to create effective refractive indices without using dielectrics. This property was used to design lenses [14] or even prisms to compensate the dispersion of a conventional leaky wave antenna [15]. For this type of application, the structure of the pins is well located within the parallel plate structure and modifies the dispersion characteristics of the first propagating mode.

Recently, higher symmetries have been proposed for periodic structures in the form of corrugations [13,16] to obtain either less dispersive behavior of the propagating mode for designing lenses [17] or to increase the stopbands for use in gap waveguide technology. Other similar periodic structures have been studied and applied in the optical domain [18,19].

The purpose of this work is to present a complete analysis of pin structures inside a parallel plate waveguide in terms of stopbands, equivalent refractive index and dispersion of the first propagating mode. In the study, we focus on the analysis of innovative configurations of pins in terms of a combination of pins with different characteristics or the use of higher symmetries such as glide symmetry.

## 2. Geometries Considered in the Analysis

The pin structures that are considered in this study are presented in Figure 1.

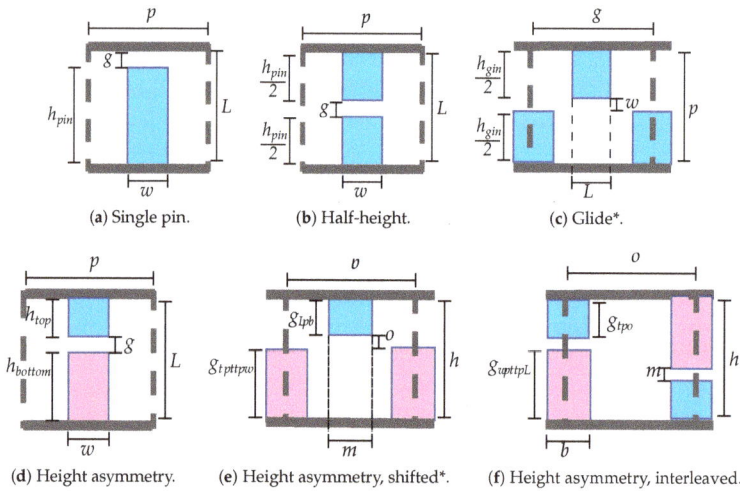

**Figure 1.** Pin geometries considered in the study. * In these cases, we will consider two cases of shift in one direction ($X$) and two directions ($XY$) shifts.

In our work, the reference case is the conventional pin (Figure 1a), where $h_{pin}$ is the height of the pin, $w$ is the width of the pin, $L$ is the distance between the two plates and $g$ is the gap either between the top plate and the pin or between the top and bottom pins. A second case of the study is a symmetric double pin with half of the height of the reference pin ($h_{pin}/2$) as shown in Figure 2b. This geometry has been extensively studied in the paper [11] in terms of its stopband properties. We will also include here the propagation properties, i.e., the dispersion of the first mode. A third case is the glide symmetric case (Figure 2c), i.e., a translation in the longitudinal direction of $p/2$ and a reflection with respect to the $L_{PPWG}/2$ plane. Here, we consider the cases of only the half-pin translation in one direction (generically named as $X$) and the shift in two directions (named as $XY$).

Three other structures are proposed in this study and are described in Figure 1d–f. The key difference of these structures is that they present asymmetries related to the previous ones already introduced. In particular, we will study the effect of using unequal pins in the top (with height $h_{top}$) and bottom (with height $h_{bottom}$) of the metal plate and also interleaving these different pins. Finally, Figure 2 contains additional representations of the cases under study including the reference coordinate axes.

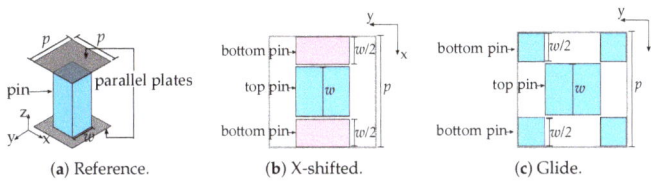

| (a) Reference. | (b) X-shifted. | (c) Glide. |

**Figure 2.** Unit cell description including axes. Top view of the translations in only X (**b**) or XY (**c**) directions.

In order to compare the six structures, some key parameters will be fixed for all the cases. The distance $L$ between the two plates, the gap size $g$, the pin width $w$ and the periodicity $p$ will remain constant in this initial study. The effect of all these parameters has been extensively analyzed in [4] for the reference case. In the second part of this work, we also include a parametric study for all the other proposed geometries.

## 3. Stopband Analysis

Simulations of the parallel plate stopband were carried out by calculating the dispersion diagram of a unit cell using the commercial software CST Microwave Studio. Alternative methodologies for calculating dispersion diagrams of these kinds of structures were presented in [20].

As for the pin structure, we have selected the following dimensions: $h_{pin} = 2.5$ mm, $w = 1$ mm, $p = 2$ mm and $g = 0.5$ mm. However, we have decided to normalize all the frequencies to the frequency $f_0$ where the pin height for the reference case is $0.25\lambda$. As a consequence and without loss of generality, with the selected $h_{pin}$ of 2.5 mm, $f_0$ corresponds to 30 GHz.

The selection of some initial dimensions and their normalization with respect to a frequency is a common practice when doing parametric studies(see the previous examples in [4,11,12]). It is important to mention that as a consequence of the linearity of the electromagnetism, the results can be directly scaled to any desired frequency range.

First, the calculated dispersion diagrams of the first two modes for the reference pin (Figure 1a), the double pin (Figure 2b) and glide symmetry cases with a shift in one direction and two directions (Figure 2c) are presented in Figure 3. In this figure the blue curves correspond to the dispersion diagram of the first mode of the geometry while the red curves correspond to the second propagating mode of the geometry. We can observe that reference (case *a*) has the largest stopband. The case named half-height (case *b*) has a higher cutoff frequency of the first mode, and the second mode is flat in frequency, meaning that it cannot be used to propagate energy and corresponds to a resonance. Finally, in the glide case, when we shift the pins (case *c*) in the bottom layer by half the period with respect to the pins in the top layer (in one or two directions), the stopband disappears between the first and second modes, as previously observed for glide corrugations [16]. Note that in this case there is a very narrow stopband between the second and third modes.

Concerning the analysis of the asymmetrical cases from Figure 1, we start with case *d* (Figure 1d). In Figure 4, the variation in the dispersion diagram for the double-pin structure when the pins in the top and bottom plates have different heights ($h_{top}$ and $h_{bottom}$, respectively) is represented. For this analysis, we always consider that the total height ($h_{top} + h_{bottom}$) is equal to the height of the reference pin $h_{pin} = 2.5$ mm. The consequence of increasing the height of one of the pins is a reduction in the cutoff frequency of the first and second modes.

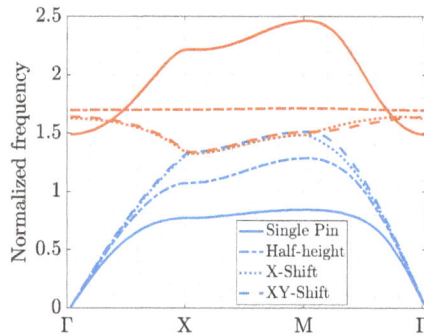

**Figure 3.** Dispersion diagrams for cases *a, b* and *c* according to Figure 1. The frequency is normalized to the frequency corresponding to the single-pin height $h_{pin} = \lambda/4$. Blue lines correspond to the first mode and red lines to the second mode.

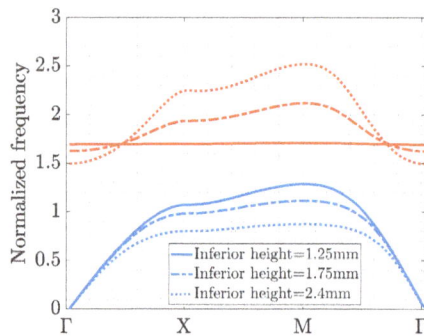

**Figure 4.** Dispersion diagrams for case *d* in Figure 1 for different pin heights, where $h_{top} + h_{bottom} = h_{pin}$. Blue lines represent the first mode and red lines the second mode.

The next step was to analyze the effect of shifting the asymmetrical pins (case *e*) as described in Figure 1e. An example of this study is presented in Figure 5. First, we considered a shift in one direction (Figure 5a) and then a shift in XY (Figure 5b). It can be seen that the effects in both cases are similar. Only when the two pins (top and bottom) had identical heights, the stopband between the first two modes disappeared. When the pins had different heights, the stopband existed.

The last case to be analyzed was the interleaving of pins with different heights in the top and bottom layers (case *f*) as shown in Figure 1f. In this particular case, we needed to use a unit cell with a doubled size to ensure the right symmetry when applying periodic boundary conditions. The results for two different interleaved heights ($h_{1_{bottom}} = 1.75$ mm and $h_{2_{bottom}} = 2.4$ mm) are shown in Figure 6. In this case, a narrow stopband occurs only for very-high-order modes. We can conclude for this case that the high number of modes is just a consequence of the supersymmetry of the unit cell. Moreover, we can easily see how there is continuity among the first five modes, which in practice is just a single mode. Additionally, note that the stopband is in the same frequency range as that for the previously studied geometries.

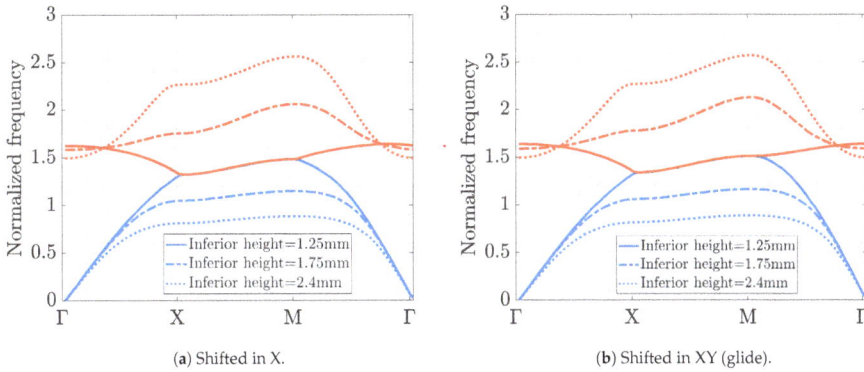

(a) Shifted in X.

(b) Shifted in XY (glide).

**Figure 5.** Dispersion diagrams for case *e* in Figure 1 for different pin heights, where $h_{top} + h_{bottom} = h_{pin}$. Blue lines represent the first mode and red lines the second mode.

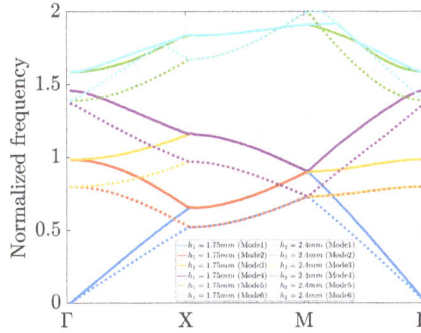

**Figure 6.** Dispersion diagrams for pins in case *f* according to Figure 1. Different colors represent different modes.

Finally, a summary of the obtained stopbands for the different structures is represented in Table 1, with $f_{start}$ corresponding to the normalized starting frequency of the stopband and $f_{stop}$ being the normalized end frequency of the stopband. The cases with glide symmetry (case *c*) are not included as the stopband is practically nonexistent. The case of interleaved pins refers to the bandgap between high-order modes; nonetheless, this stopband exists in the same frequency range as that for the other geometries.

**Table 1.** Summary of the starting and end frequencies of the stopbands created by the different analyzed structures.

| Geometry | $f_{start}$ | $f_{end}$ |
|---|---|---|
| Single pin (case *a*) | $0.84f_0$ | $1.5f_0$ |
| Half-pin (case *b*) | $1.28f_0$ | $1.7f_0$ |
| Pins, different heights (case *d*, $h_1 = 1.5$) | $1.23f_0$ | $1.68f_0$ |
| Pins, different heights (case *d*, $h_1 = 2$) | $1.01f_0$ | $1.58f_0$ |
| Pins, shifted (case *e*, $h_1 = 1.5$) | $1.29f_0$ | $1.51f_0$ |
| Pins, shifted (case *e*, $h_1 = 2$) | $1.03f_0$ | $1.55f_0$ |
| Pins, glide (case *e*, $h_1 = 1.5$) | $1.31f_0$ | $1.53f_0$ |
| Pins, glide (case *e*, $h_1 = 2$) | $1.04f_0$ | $1.55f_0$ |
| Interleaved pins (case *f*, $h_1 = 1.5$) | $1.31f_0$ | $1.68f_0$ |
| Interleaved pins (case *f*, $h_1 = 2$) | $1.38f_0$ | $1.57f_0$ |

From this first comparative analysis among the different proposed geometries, several conclusions are derived. When dealing with shifted geometries (X or XY), the use of unequal pins in the top and bottom plates preserved the stopbands, whilst the use of identical pins made it disappear. On the other hand, when these unequal pins were interleaved, again there was no stopband.

*Parametric Study for Cases with Pins of Unequal Heights*

In this section, we include a brief parametric study of some of the geometries. The type of representation is similar to the ones presented in [4,11]. The complete study of all the cases in most situations gives the same type of conclusions as in the previous studies, and for this reason we have selected some representative cases to be shown here. The range of variation of the different parameters has been also taken from these previous studies as the intention of the present work is to analyze the effect of the specific geometry.

The first geometry to be analyzed is the the case of double pins with unequal heights, denoted as case *d* (Figure 1d). The parameters that vary are the gap *g* for different periodicities and the effect of the relation $w/p$ between the width *w* of the pin and the period *p* for different gap sizes. These parameters are analyzed for pin heights values $h_{bottom}$ of 1.5 mm and 2 mm, and the results are shown in Figures 7 and 8, respectively. In these figures, the blue curves represent the upper limit of the stopband, while the red curves represent the lower limit. For each value of the horizontal axis, a dispersion diagram has been calculated. The two represented points (red and blue) in vertical, correspond to the frequency where the stopband stats and the frequency where the stopband stops. For example, if you consider the dispersion diagram in Figure 3 there is a frequency range where there are no modes. The maximum frequency of propagation of the first mode is the lower limit of the stop band, and the minimum frequency of propagation of the second mode is the upper limit.

The frequency is normalized to $f_0$, and the other parameters are normalized to $\lambda_0$. The parameters that are not changed in the analysis correspond to the reference case. For the structure made with double pins of unequal heights, a reduction in the gap size always increases the stopband as shown in Figures 7a and 8a. As the periodicity increases, the lower limit of the stopband decreases, and the upper limit of the stopband is not affected up to a value of the period after which it decreases. The ratio between the width of the pin and the periodicity $w/p$ is optimum (maximizing the stopband) when the proportion is 1 to 2 (Figures 7b and 8b).

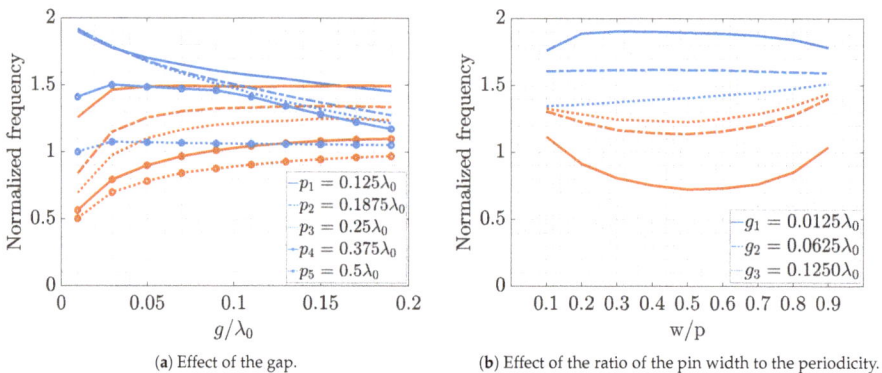

(**a**) Effect of the gap.  (**b**) Effect of the ratio of the pin width to the periodicity.

**Figure 7.** Parametric study for the geometry described as *d* in Figure 1 for $h_{bottom} = 1.5$ mm. Red lines represent the start frequency of the stopband whilst blue lines represent its end frequency.

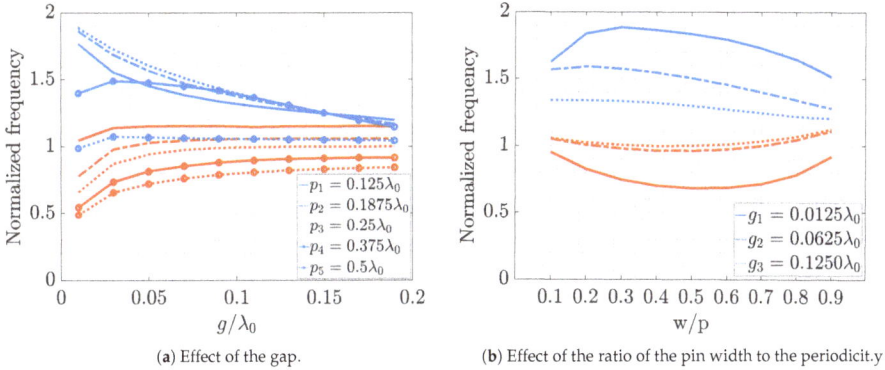

(a) Effect of the gap.

(b) Effect of the ratio of the pin width to the periodicit.y

**Figure 8.** Parametric study for the geometry described as $d$ in Figure 1 for $h_{bottom} = 2.0$ mm. Red lines represent the start frequency of the stopband whilst blue lines represent its end frequency.

For the case of geometries with shifted pins (case $e$), similar studies were carried out. In this case, we show the parametric effect when the pins have the same height and different heights. Two cases have been considered: pins shifted in just one direction (Figures 9 and 10) and pins shifted in both directions (Figures 11 and 12). In these figures, again the red lines represent the lower limit of the stopband, and the blue lines represent the upper limit. In addition, we have made use of subplots to better show the variation of the parameters. In these subplots, the dashed lines correspond to the stopbands formed between the second and third modes (represented by shaded areas), while the solid lines correspond to the stopband formed between the first and second modes. As concluded before, these geometries have very narrow or even nonexistent stopbands when $h_{top} = h_{bottom}$. However, when the pins have different sizes, stopbands appear and follow similar rules in terms of their parametric dependence as in the non shifted cases. Finally, the differences between the shifted cases in just one direction and in two directions are very small for the case when the pins have unequal heights but more important for the case when the pins have the same height (glide case).

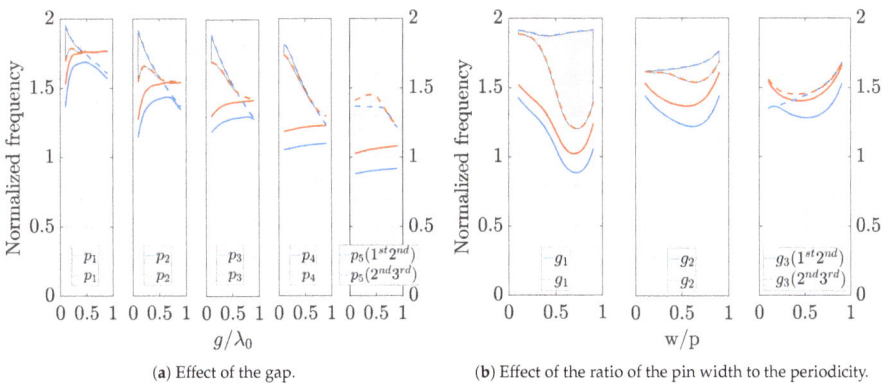

(a) Effect of the gap.

(b) Effect of the ratio of the pin width to the periodicity.

**Figure 9.** Parametric study for the X-shifted geometry with $h_{top} = h_{bottom}$, $p_1 = 0.125\lambda_0$, $p_2 = 0.1875\lambda_0$, $p_3 = 0.25\lambda_0$, $p_4 = 0.375\lambda_0$, $p_5 = 0.5\lambda_0$, $g_1 = 0.0125\lambda_0$, $g_2 = 0.0625\lambda_0$, and $g_3 = 0.125\lambda_0$.

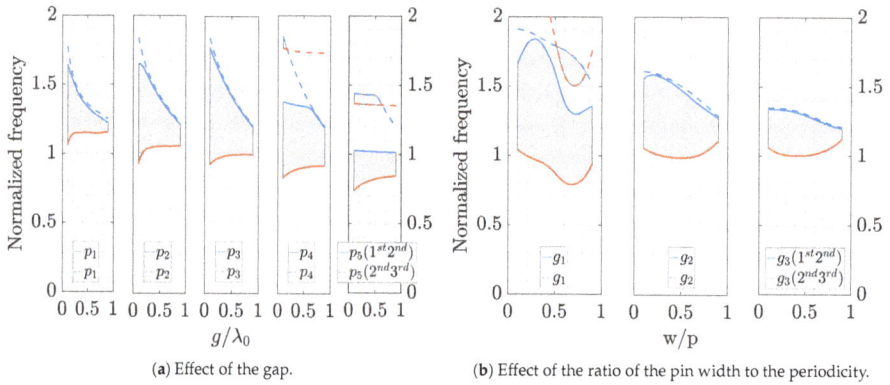

(a) Effect of the gap.

(b) Effect of the ratio of the pin width to the periodicity.

**Figure 10.** Parametric study for the X-shifted geometry with $h_{bottom} = 2$ mm, $p_1 = 0.125\lambda_0$, $p_2 = 0.1875\lambda_0$, $p_3 = 0.25\lambda_0$, $p_4 = 0.375\lambda_0$, $p_5 = 0.5\lambda_0$, $g_1 = 0.0125\lambda_0$, $g_2 = 0.0625\lambda_0$, and $g_3 = 0.125\lambda_0$.

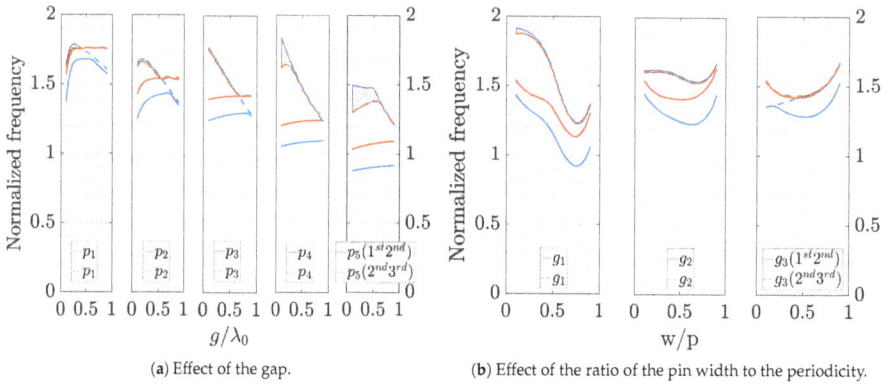

(a) Effect of the gap.

(b) Effect of the ratio of the pin width to the periodicity.

**Figure 11.** Parametric study for the XY-shifted geometry with $h_{top} = h_{bottom}$, $p_1 = 0.125\lambda_0$, $p_2 = 0.1875\lambda_0$, $p_3 = 0.25\lambda_0$, $p_4 = 0.375\lambda_0$, $p_5 = 0.5\lambda_0$, $g_1 = 0.0125\lambda_0$, $g_2 = 0.0625\lambda_0$, and $g_3 = 0.125\lambda_0$.

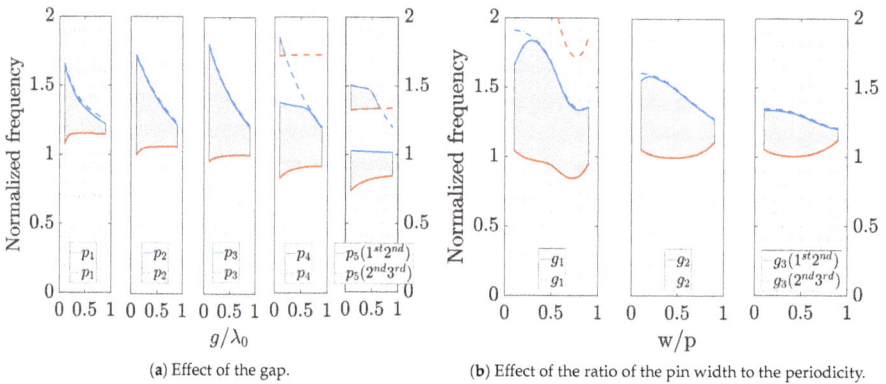

(a) Effect of the gap.

(b) Effect of the ratio of the pin width to the periodicity.

**Figure 12.** Parametric study for the XY-shifted geometry with $h_{bottom} = 2$ mm, $p_1 = 0.125\lambda_0$, $p_2 = 0.1875\lambda_0$, $p_3 = 0.25\lambda_0$, $p_4 = 0.375\lambda_0$, $p_5 = 0.5\lambda_0$, $g_1 = 0.0125\lambda_0$, $g_2 = 0.0625\lambda_0$, and $g_3 = 0.125\lambda_0$.

## 4. Equivalent Refractive Index

Now, we proceed to analyze the effective refractive index of the first propagating mode of the previous structures. There are two main parameters we need to consider, namely the range of values achieved by the different structures (as all of them have the same basic element and periodicity) and how flat these lines are in frequency, as this will measure the dispersion of the element. The results can be used for designing wideband lenses [21], making use of the low dispersion characteristic, or conversely to design a structure that requires a very dispersive unit cell as in [15].

The calculation of the equivalent refraction index is made from the calculated dispersion diagram. In this calculation, we obtain the phase constant ($\beta$) of the first propagating mode. By dividing this phase constant by the free space propagation constant ($\kappa_0$), we obtain the equivalent refractive index.

Figure 13a shows the effect of using the structures made with double pins with different heights in comparison with the case of a single pin. It is clear that with a single pin, we obtain higher refraction indices, but at the same time the structure is much more dispersive. Note that the less dispersive case occurs when both pins have exactly the same height.In all these figures the lines stop in the vertical axis due to the beginning of the stopband.

Figure 13b shows the analysis of the case with interleaving pins (case $f$ of Figure 1). Once again, different heights of the pins have been considered. It is observed that changing the sizes of the consecutive pins leads to considerably higher equivalent refractive indices. This is explained by the fact that this is the only case where the air gap in between pins is not always located in the same position in the cross-section of the structure, as the gap position changes between consecutive pins. This somehow has an effect on the propagating field such that it does not encounter a straight path with only air. On the contrary, the propagating field necessarily finds the pins in its way, consequently increasing the equivalent refractive index. Another observation from these results is that with the interleaving pins, the mode is less dispersive.

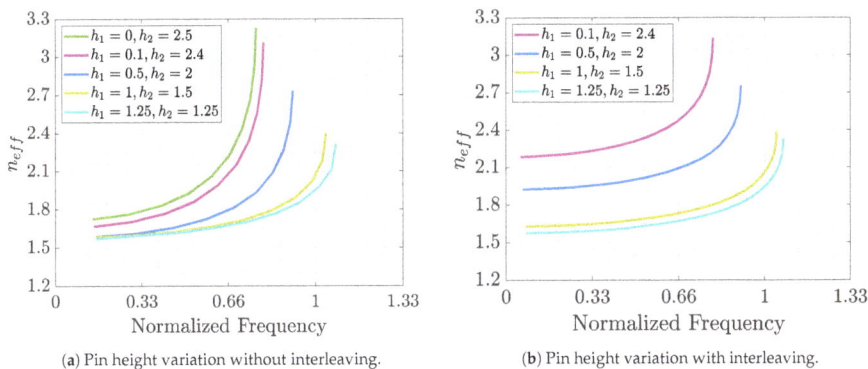

(a) Pin height variation without interleaving.

(b) Pin height variation with interleaving.

**Figure 13.** Equivalent refractive indices for different cases from Figure 1: in particular cases, $a$, $b$, $d$ and $f$.

In Figure 14, we show the effect of using higher symmetries in the unit cells of cases $b$ and $d$ of Figure 1. For case $b$ (identical pins on the top and bottom), we can clearly see in Figure 14a that a displacement of half the periodicity in one direction makes the structure less dispersive, as previously demonstrated for other types of unit cells [12]. An extra displacement added in the perpendicular direction shas a small effect in this case, probably due to the fact that we are calculating the equivalent refractive index only in the first zone of the Brillouin diagram ($\Gamma$-$X$). The achieved equivalent refraction indices are lower for the case with glide symmetry than for the case of the original pins without shifting. The same effect is observed for the case in which the top and bottom pins have different heights (Figure 14b).

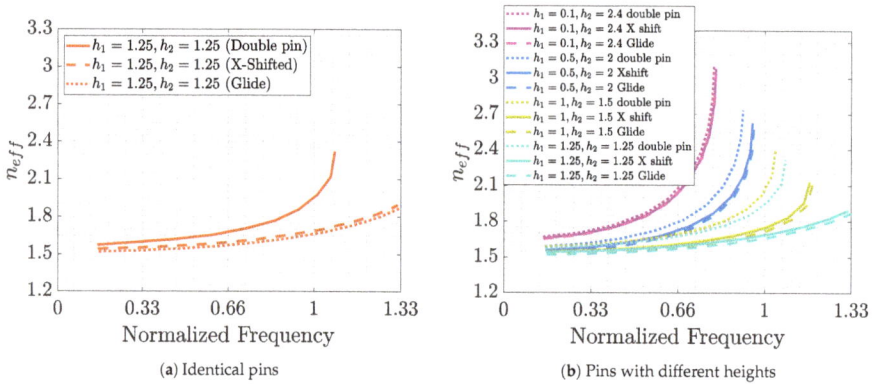

(a) Identical pins

(b) Pins with different heights

**Figure 14.** Equivalent refractive indices for different cases in Figure 1 after shifting the pins.

*Effect of the Different Parameters on the Equivalent Refractive Index*

In this section, we present the effect of varying some of the parameters of the analyzed structures on the equivalent refractive index and the dispersion of the first mode. For this analysis, we individually change the size of the gap, the width of the pin and the periodicity, while the other parameters are kept constant at the initial reference values.

The first case to study is the structure with double pins of different heights. The reference case variables values in this case are $g = 0.5$ mm, $p = 2$ mm and $w = 1$ mm. The results are presented in Figure 15. We can observe the following general conclusions: a decrease in the gap to one half of the value of the reference case (Figure 15b) produces an increase in the equivalent refraction index for all the cases, probably as a consequence of the increase in the amount of metal with respect to air in the unit cell. On the other hand, an increase in the periodicity (from 2 to 3 mm) decreases the equivalent refractive index and makes the mode slightly less dispersive (Figure 15c). Finally, the increase in the width of the pins to 1.5 mm decreases the minimum value of the refraction indices for all the cases independently of the heights (shown in Figure 15d).

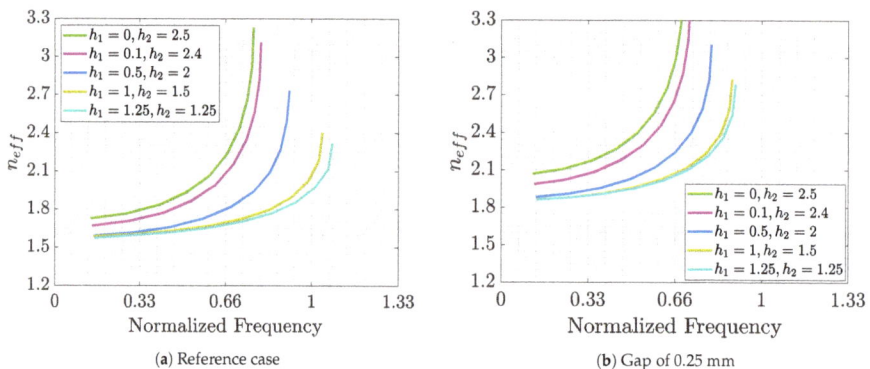

(a) Reference case

(b) Gap of 0.25 mm

**Figure 15.** *Cont.*

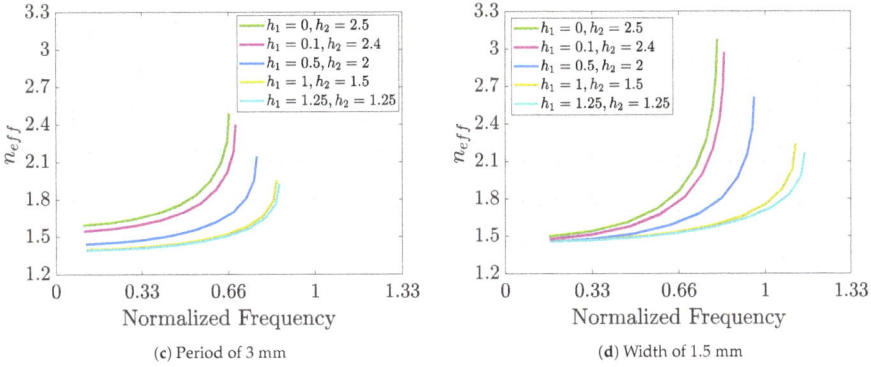

(**c**) Period of 3 mm

(**d**) Width of 1.5 mm

**Figure 15.** Effect of varying different parameters in the geometries *a*, *b* and *d* from Figure 1. The reference case has a gap of 0.5 mm, a period of 2 mm and a width of 1 mm.

Considering the case *f* from Figure 1, i.e., interleaving pins with different heights, the effect of the parameters is shown in Figure 16. In this case, a change in the gap size produces the same effects as those in the previous studied case. The use of a higher period is useful to obtain smaller values of the equivalent refraction indices but in this case does not show an improvement in the dispersion of the mode. Finally, the increase in the width of the pins reduces the dispersion in all the cases and has different consequences for the achieved values of the refraction indices depending on the heights. For identical pin heights, there is a reduction in the equivalent refractive index, while for unequal pins the index increases.

When the top and bottom pins are shifted to enable higher symmetries, the effect of varying the same parameters is presented in Figure 17 for the case using pins with identical heights (glide case). The conclusions for this case are similar to those for the previous cases in terms of the variation in the gap and the periodicity (decreasing the gap or decreasing the periodicity increases the equivalent refractive index). On the other hand, the structure is almost insensitive to a variation in the width of the pins.

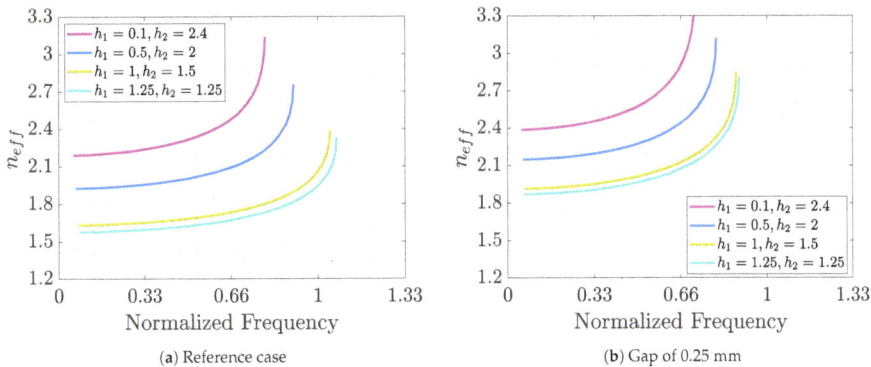

(**a**) Reference case

(**b**) Gap of 0.25 mm

**Figure 16.** *Cont.*

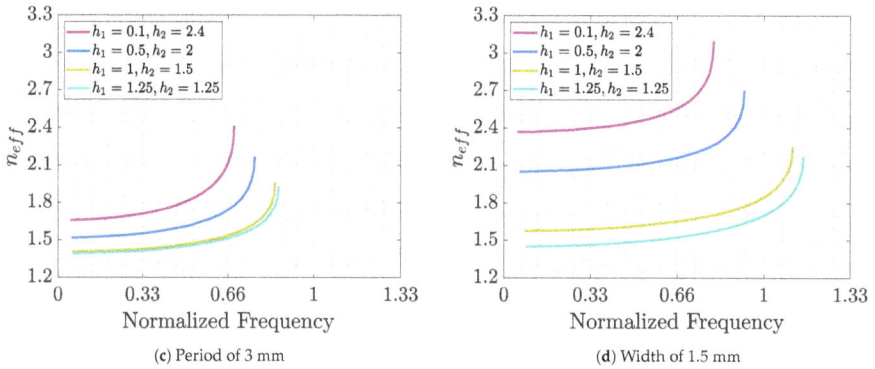

**Figure 16.** Effect of varying different parameters in the geometry $f$ (interleaved pins) from Figure 1. The reference case has a gap of 0.5 mm, a period of 4 mm and a width of 1 mm.

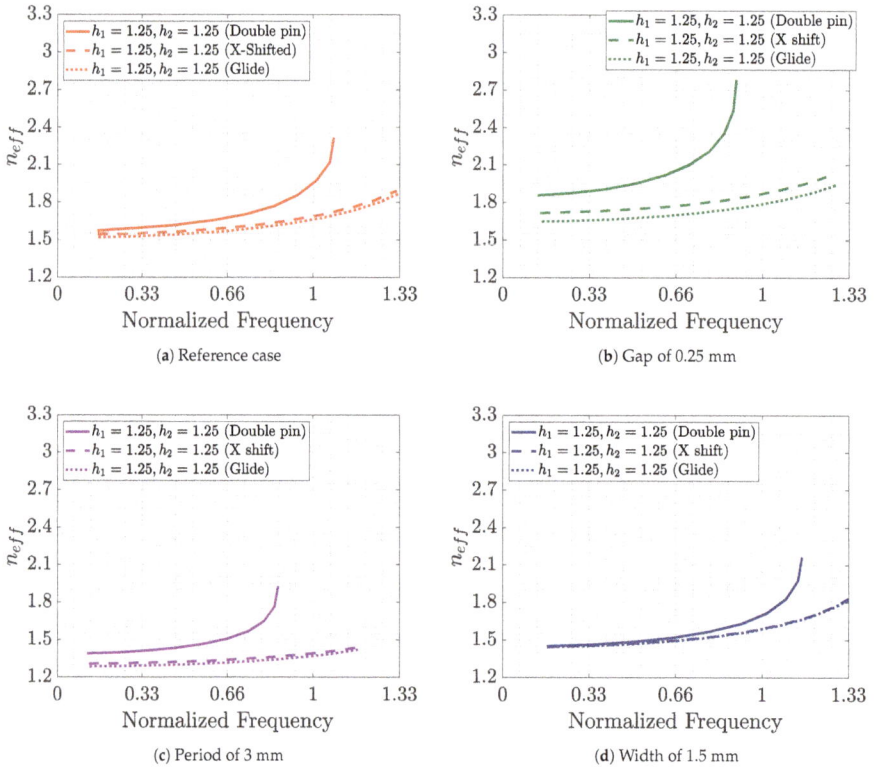

**Figure 17.** Effect of varying different parameters in the geometry $c$ from Figure 1. The reference case has a gap of 0.5 mm, a period of 4 mm and a width of 1 mm.

Finally, in the latter case, when we analyzed the behavior when the pins have unequal heights in the top and bottom layers, i.e., case $e$ from Figure 1, the most significant conclusion derived from the results shown in Figure 18 is the insensitivity to the variation in the width of the pins. In fact, with the increase in the width of the pins from 1 mm to 1.5 mm, all the analyzed cases converged to the same value of equivalent refractive index at low frequencies independently of their height or shift. On the other hand, for the other two parameters (the gap size and period), the conclusions were not different from those obtained for the previous cases.

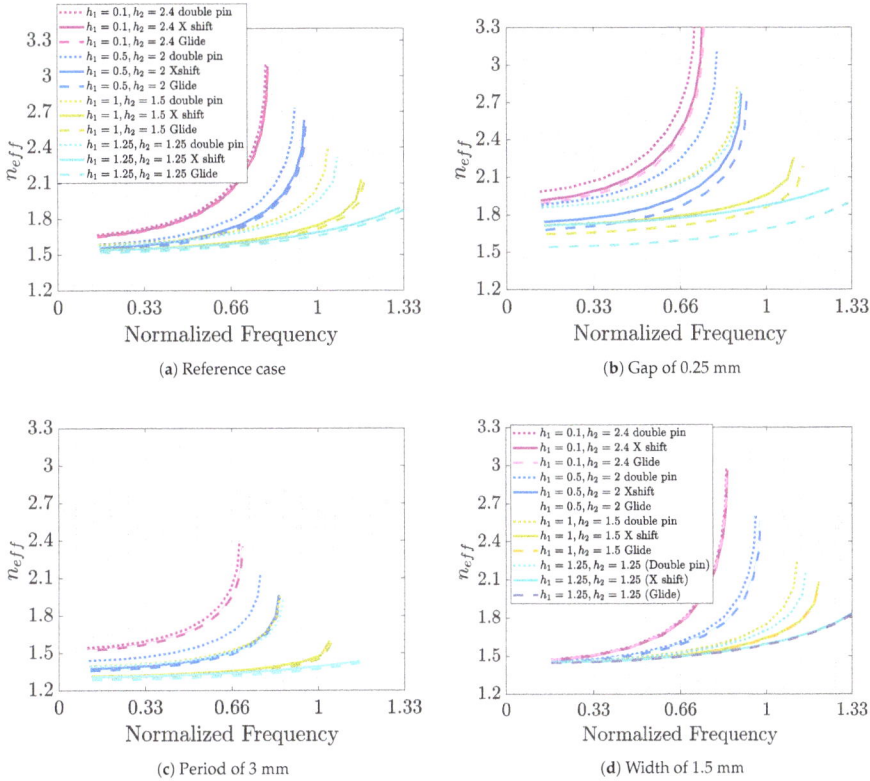

**Figure 18.** Effect of varying different parameters in the geometry *e* from Figure 1. The reference case has a gap of 0.5 mm, a period of 4 mm and a width of 1 mm.

## 5. Conclusions

In this work, different periodic structures combining pins inside parallel plate waveguide structures are analyzed. The pins are realized in several configurations including glide symmetry, shifting of the pins, asymmetries in the pin height and interleaving pins with different heights. The study focuses on dispersion diagrams. From the computed dispersion diagrams, information about the existence of a stopband for the parallel plate modes and its size and the propagation characteristics of the first mode are obtained. The propagation characteristics are studied as equivalent refraction indices as a function of frequency.

Initially, the effect of the different geometries considered (presented in Figure 1) in these two aspects are studied. Concerning the stopband, the reference geometry (a single pin) produces a stopband covering a larger frequency range. On the other hand, glide symmetric structures do not have a stopband, while the cases combining pins with unequal heights always have stopbands (even if they are narrow).

The parametric studies carried out regarding this property confirm that in all cases with stopbands, a reduction in the gap size increases the bandwidth of the stopband and the ratio between the pin

*Symmetry* **2019**, *11*, 582

width and the period is relevant in determining this bandwidth. Similar conclusions were obtained in the past for the reference case.

With respect to the effective refractive index of the first propagating transverse electromagnetic mode (TEM), the main conclusions are that the structure with glide symmetry is less dispersive and that even for the case of using top and bottom pins with different heights, applying half-period shifts always reduces the dispersion. On the other side, the most dispersive case is the reference case with a single pin. Finally, when the pins are interleaved, the mode is also less dispersive than the reference structure and the equivalent refractive indices are larger.

The effect of the different parameters has also been evaluated from the perspective of propagation characteristics. As a global conclusion, a reduction in the gap or the period increases the refraction index. The effect of the width of the pins depends on the specific geometry.

Structures made with pins have been experimentally used and the manufacturing process is typically either a milling or 3D printing. When moving far up in frequency, long and thin pins can result in too fragile for manufacturing and the presented geometries based on the use of shorter pins can have advantages from this point of view.

**Author Contributions:** The contributions of all of the authors were the same. All of authors worked together to develop the present manuscript.

**Funding:** This research was funded by [the Spanish Ministerio de Ciencia, Innovación y Universidades grant number [TEC2016-79700-C2-2-R], Becas Santander Jóvenes Investigadores and FONDECYT 11180434 project.

**Conflicts of Interest:** The authors declare no conflict of interest.

## Abbreviations

The following abbreviations are used in this manuscript:

CST     Computer Simulation Studio
PPWG   Parallel plate waveguide
AMC    Artificial magnetic conductor
TEM    Transverse electromagnetic mode

## References

1. Silveirinha, M.G.; Fernandes, C.A.; Costa, J.R. Electromagnetic Characterization of Textured Surfaces Formed by Metallic Pins. *IEEE Trans. Antennas Propag.* **2008**, *56*, 405–415. [CrossRef]
2. Kildal, P.; Alfonso, E.; Valero-Nogueira, A.; Rajo-Iglesias, E. Local Metamaterial-Based Waveguides in Gaps Between Parallel Metal Plates. *IEEE Antennas Wirel. Propag. Lett.* **2009**, *8*, 84–87. [CrossRef]
3. Kildal, P.; Zaman, A.U.; Rajo-Iglesias, E.; Alfonso, E.; Valero-Nogueira, A. Design and experimental verification of ridge gap waveguide in bed of nails for parallel-plate mode suppression. *IET Microw. Antennas Propag.* **2011**, *5*, 262–270. [CrossRef]
4. Rajo-Iglesias, E.; Kildal, P. Numerical studies of bandwidth of parallel-plate cut-off realised by a bed of nails, corrugations and mushroom-type electromagnetic bandgap for use in gap waveguides. *IET Microw. Antennas Propag.* **2011**, *5*, 282–289. [CrossRef]
5. Rajo-Iglesias, E.; Zaman, A.U.; Kildal, P. Parallel Plate Cavity Mode Suppression in Microstrip Circuit Packages Using a Lid of Nails. *IEEE Microw. Wirel. Compon. Lett.* **2010**, *20*, 31–33. [CrossRef]
6. Pucci, E.; Rajo-Iglesias, E.; Kildal, P. New Microstrip Gap Waveguide on Mushroom-Type EBG for Packaging of Microwave Components. *IEEE Microw. Wirel. Compon. Lett.* **2012**, *22*, 129–131. [CrossRef]
7. Rajo-Iglesias, E.; Kildal, P.; Zaman, A.U.; Kishk, A. Bed of Springs for Packaging of Microstrip Circuits in the Microwave Frequency Range. *IEEE Trans. Compon. Packag. Manuf. Technol.* **2012**, *2*, 1623–1628. [CrossRef]
8. Rajo-Iglesias, E.; Pucci, E.; Kishk, A.A.; Kildal, P. Suppression of Parallel Plate Modes in Low Frequency Microstrip Circuit Packages Using Lid of Printed Zigzag Wires. *IEEE Microw. Wirel. Compon. Lett.* **2013**, *23*, 359–361. [CrossRef]

9.  Sánchez-Escuderos, D.; Ferrando-Bataller, M.; Berenguer, A.; Baquero-Escudero, M.; Valero-Nogueira, A. Dielectric Bed of Nails in Gap-Waveguide Technology at Millimeter-Wave Frequencies. *IEEE Microw. Wirel. Compon. Lett.* **2014**, *24*, 515–517. [CrossRef]

10. Rajo-Iglesias, E.; Ferrando-Rocher, M.; Zaman, A.U. Gap Waveguide Technology for Millimeter-Wave Antenna Systems. *IEEE Commun. Mag.* **2018**, *56*, 14–20. [CrossRef]

11. Fan, F.; Yang, J.; Vassilev, V.; Zaman, A.U. Bandwidth Investigation on Half-Height Pin in Ridge Gap Waveguide. *IEEE Trans. Microw. Theory Tech.* **2018**, *66*, 100–108. [CrossRef]

12. Ebrahimpouri, M.; Quevedo-Teruel, O.; Rajo-Iglesias, E. Design Guidelines for Gap Waveguide Technology Based on Glide-Symmetric Holey Structures. *IEEE Microw. Wirel. Compon. Lett.* **2017**, *27*, 542–544. [CrossRef]

13. Ebrahimpouri, M.; Rajo-Iglesias, E.; Sipus, Z.; Quevedo-Teruel, O. Cost-Effective Gap Waveguide Technology Based on Glide-Symmetric Holey EBG Structures. *IEEE Trans. Microw. Theory Tech.* **2018**, *66*, 927–934. [CrossRef]

14. Maci, S.; Minatti, G.; Casaletti, M.; Bosiljevac, M. Metasurfing: Addressing Waves on Impenetrable Metasurfaces. *IEEE Antennas Wirel. Propag. Lett.* **2011**, *10*, 1499–1502. [CrossRef]

15. Wang, L.; Gómez-Tornero, J.L.; Rajo-Iglesias, E.; Quevedo-Teruel, O. Low-Dispersive Leaky-Wave Antenna Integrated in Groove Gap Waveguide Technology. *IEEE Trans. Antennas Propag.* **2018**, *66*, 5727–5736. [CrossRef]

16. Valerio, G.; Sipus, Z.; Grbic, A.; Quevedo-Teruel, O. Accurate Equivalent-Circuit Descriptions of Thin Glide-Symmetric Corrugated Metasurfaces. *IEEE Trans. Antennas Propag.* **2017**, *65*, 2695–2700. [CrossRef]

17. Quevedo-Teruel, O.; Miao, J.; Mattsson, M.; Algaba-Brazalez, A.; Johansson, M.; Manholm, L. Glide-Symmetric Fully Metallic Luneburg Lens for 5G Communications at Ka-Band. *IEEE Antennas Wirel. Propag. Lett.* **2018**, *17*, 1588–1592. [CrossRef]

18. Mendis, R.; Mittleman, G. Artificial Dielectrics: Ordinary Metallic Waveguides Mimic Extraordinary Dielectric Media. *IEEE Microw. Mag.* **2014**, *15*, 32–45. [CrossRef]

19. Chen, L.; Cheng, Z.; Xu, J.; Zanf, X.; Cai, B.; Zhu, Y. Controllable multiband terahertz notch filter based on a parallel plate waveguide with a single deep groove. *Opt. Lett.* **2014**, *39*, 4541–4544. [CrossRef]

20. Mesa, F.; Rodríguez-Berral, R.; Medina, F. On the Computation of the Dispersion Diagram of Symmetric One-Dimensionally Periodic Structures. *Symmetry* **2018**, *10*. [CrossRef]

21. Quevedo-Teruel, O.; Ebrahimpouri, M.; Ng Mou Kehn, M. Ultrawideband Metasurface Lenses Based on Off-Shifted Opposite Layers. *IEEE Antennas Wirel. Propag. Lett.* **2016**, *15*, 484–487. [CrossRef]

*symmetry*

MDPI

*Article*

# One-Plane Glide-Symmetric Holey Structures for Stop-Band and Refraction Index Reconfiguration

Adrian Tamayo-Dominguez [1,*], Jose-Manuel Fernandez-Gonzalez [1] and Oscar Quevedo-Teruel [2]

[1]  Department of Signals, Systems and Radiocommunications, Universidad Politécnica de Madrid, 28040 Madrid, Spain; jmfdez@gr.ssr.upm.es
[2]  Division of Electromagnetic Engineering, KTH Royal Institute of Technology, 10044 Stockholm, Sweden; oscarqt@kth.se
*   Correspondence: a.tamayo@upm.es

Received: 1 March 2019; Accepted: 1 April 2019; Published: 4 April 2019

**Abstract:** This work presents a new configuration to create glide-symmetric structures in a single plane, which facilitates fabrication and avoids alignment problems in the assembly process compared to traditional glide-symmetric structures based on several planes. The proposed structures can be printed on the metal face of a dielectric substrate, which acts as a support. The article includes a parametric study based on dispersion diagrams on the appearance of stop-bands and phase-shifting by breaking the symmetry. In addition, a procedure to regenerate symmetry is proposed that may be useful for reconfigurable devices. Finally, the measured and simulated S parameters of 10 × 10 unit-cell structures are presented to illustrate the attenuation in these stop-bands and the refractive index of the propagation modes. The attenuation obtained is greater than 30 dB in the stop-band for the symmetry-broken prototype.

**Keywords:** glide symmetry; single plane; stop-band; periodic structures; higher symmetries; refractive index

---

## 1. Introduction

The recent study of higher symmetries [1–5] was driven by a growing interest in the use of periodic structures to improve the electromagnetic properties of antennas and microwave devices. These symmetries were first investigated in the 1960s and 1970s for one-dimensional periodic structures [6–8], introducing the concepts of glide and screw (twist) symmetry. More recently, two-dimensional glide symmetries were proposed and studied, which are a particular case of higher symmetries, demonstrating great potential for modifying the dispersion properties of periodic structures [9,10]. Periodic glide-symmetric structures are obtained by translating and mirroring a unit cell with respect to the glide plane [11]. The theoretical analysis of some of these structures with glide symmetry was carried out using the Floquet theorem [12–14], which provides an effective tool for the analysis of periodic structures. Glide symmetries were successfully used to reduce the dispersion of periodic structures [15–18], to increase the equivalent refractive index [19–22], or to increase the band and attenuation of electromagnetic bandgaps [10,23–25]. For example, glide symmetry was proposed to produce lens antennas for fifth-generation (5G) communications [26,27], taking advantage of their ability to generate a higher refractive index, less dispersion, and more isotropy [9]. Preceding this work, all glide-symmetric structures proposed in the literature for parallel plate configurations have a horizontal plane of symmetry, perpendicular to the direction of propagation [9,10]. These configurations require two different planes that must be shifted with respect to the other, which complicates fabrication and can introduce alignment problems.

In this work, we propose a glide-symmetric structure in a single plane with ellipses as the main element on a dielectric substrate that acts as a support. Glide symmetry is achieved by placing the symmetry plane vertically, preserving the orthogonality with respect to the direction of propagation. The use of a dielectric substrate instead of an all-metal structure allows the introduction of additional degrees of freedom, including the ability to modify or break the glide symmetry in the case of using a substrate that may have different values of dielectric constant (e.g., liquid crystals [28]). The propagation takes place mainly in an air gap above the plane of the ellipses; thus, the losses that the dielectric substrate may introduce are low. The electromagnetic properties of the structure in terms of dispersion, stop-bands, or refractive index are determined by the orientation, size, and position of these ellipses.

This paper is organized in four sections. Section 2 shows the basic properties and parameters of the glide-symmetric unit cell and some examples of operation. It includes the possibility to regenerate the glide symmetry from a symmetry-broken structure, maintaining the value of all parameters. In Section 3, the performance of these new structures is illustrated by the S parameters of a complete design with 10 × 10 unit cells, comparing measurements and simulations. The extracted dispersion diagrams from measurements are compared with the simulated unit cells. Finally, Section 4 discusses the results and potential applications of the proposed structures.

## 2. Materials and Methods

The studied structures consist of two metallic layers separated by a given air gap. The top layer is a smooth metal sheet, while the bottom layer contains the glide-symmetric ellipses printed on the metal surface of a dielectric substrate. This configuration generates conditions in the electromagnetic fields inside the air gap between metal sheets that cause modifications in the propagation and the appearance of stop-bands depending on the shape and orientation of the ellipses. Figure 1 represents the unit cell with all the parameters that define its electromagnetic behavior. Figure 1a shows the thicknesses of the air $h_{Air}$ and the dielectric substrate $h_{Diel}$, together with the dielectric constant $\varepsilon_r$ of the substrate and the period $p$. Periodic conditions in the $x$- and $y$-directions of the unit cell were applied to perform the dispersion analysis. The direction of propagation selected in the analysis extends along the $y$-axis, in which there are two periodicities to generate glide symmetry. In the perpendicular direction ($x$), there is only one periodicity. Figure 1b shows the parameters related to the size, position, orientation, and shape of the ellipses, which will determine the basic behavior of the whole structure. In particular, the sizes of the major and minor semi-axes of the ellipses are defined by the values of $A_1$ and $B_1$ for the first ellipse and $A_2$ and $B_2$ for the second ellipse. The parameters $\varphi$ and $\theta$ define the angle of inclination of the first and second ellipses. A zero value means a horizontal position of the ellipses (semi-minor axis in the $y$-direction). Positive values of $\varphi$ introduce counter-clockwise rotation for the first ellipse, and positive values of $\theta$ imply clockwise rotation for the second ellipse. Finally, the distance between ellipses can be modified by the displacement parameter $d$. The ellipses are kept at a distance equal to the periodicity $p$ with $d = 0$, while values greater than zero mean that the second ellipse moved toward the first ellipse.

From the definition of the parameters presented in Figure 1, glide symmetry is achieved when $A_1 = A_2$, $B_1 = B_2$, $\varphi = \theta$, and $d = 0$. The symmetry axis is at the center of the structure along the $y$-direction, and glide symmetry is obtained by reflecting the initial ellipse through the symmetry axis and displacing the ellipse obtained at a distance $p$ in the $y$-direction. Figure 2 depicts an example of the operation of this structure. Figure 2a represents the dispersion diagrams of a structure with glide symmetry, and Figure 2b represents the same structure after breaking the symmetry with a modification of $\theta$. In this work, all the dispersion analyses were carried out using the Eigenmode solver in CST Microwave Studio considering periodic boundaries in $x$- and $y$-directions, and electrical boundaries in the $z$-direction. Although there are available methods to quickly analyze glide symmetry based on the Floquet theorem [12–14] or equivalent circuits [11,29], these methods were not proposed to analyze this structure [30]. The dispersion diagrams only represent phase variations in the $y$-direction, as it is the direction in which the glide symmetry takes place. The chosen parameters for the case shown in Figure 2a are as follows: $\varepsilon_r = 3$, $h_{Air} = 0.3$ mm, $h_{Diel} = 1.52$ mm, $p = 2.75$ mm, $A_1 = A_2 = 2.5$ mm,

$B_1 = B_2 = 0.75$ mm, $\varphi = \theta = 30°$, and $d = 0$ mm. The value of $\theta$ is changed to $90°$ in the structure of Figure 2b.

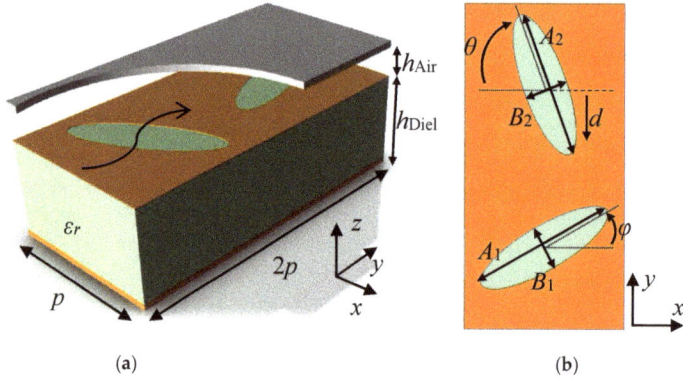

**Figure 1.** Unit cell of the proposed structure and parametrization. View in perspective (**a**), and top view without the top metal plate (**b**).

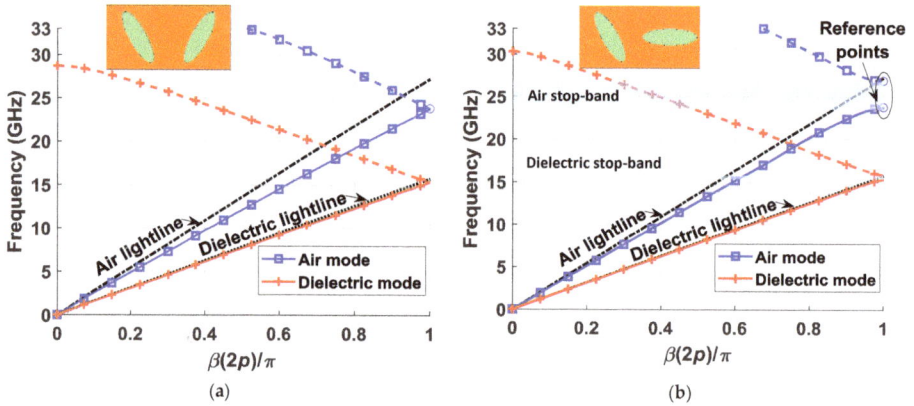

**Figure 2.** Dispersion diagrams for two configurations of the unit cell: glide-symmetric (**a**) and non-glide-symmetric (**b**).

Since the proposed structure is built in two different mediums, there are two fundamental modes of propagation, one in the air gap and one in the dielectric. The mode of interest in this case is the one that propagates in the air, as it has lower losses. An example of the dispersion curves of these modes is shown in Figure 2. For a glide configuration (Figure 2a), non-dispersive modes are obtained. However, the immediate effect of the rupture of the symmetry (Figure 2b) is the appearance of stop-bands or frequency ranges in which the fields do not propagate in the chosen direction. The stop-band appears in both the air and dielectric modes, but it is wider and its effect is more pronounced in the air mode. This is because the air gap is thinner than the thickness of the dielectric, and the electromagnetic fields confined inside are more influenced by the shape of the ellipses. The rupture of the symmetry generates a stop-band for the value $\beta(2p)/\pi = 1$ of the dispersion diagram, as expected from a periodic glide-symmetric structure with two periodicities [10]. The reference points indicated in Figure 2b represent the upper and lower limits of the stop-band, calculated for a $180°$ phase difference value between boundaries in the $y$-direction, which is equivalent to a value of 1 normalized with respect to period $2p$ of the structure. In the parametric studies carried out in Sections 2.1–2.4, these reference points are used to show the behavior of the stop-band and the refractive index. The refractive index $n$ is calculated as the ratio of the speed of light in vacuum, $c$, divided by the speed of light in the medium under study, $v$. Equations (1) and (2) demonstrate that this is equivalent to calculating the relationship between frequency in vacuum, $f_v$, and frequency in the medium under study, $f_m$, for any $\beta_v(2p)/\pi = \beta_m(2p)/\pi$ value of the dispersion diagram, assuming that periodicity $p$ is the same in both cases. Since $\beta_v(2p)/\pi = \beta_m(2p)/\pi$, we have $\lambda_v = \lambda_m$; thus, we get Equation (2).

$$n = \frac{c}{v} = \frac{\lambda_v f_v}{\lambda_m f_m};$$ (1)

$$n = \frac{f_v}{f_m}.$$ (2)

Equation (2) implies that the refractive index increases if the frequency of the mode propagating in the medium under study decreases. In the studies carried out in this section, the frequency in vacuum for the reference point $\beta(2p)/\pi = 1$, with $p = 2.75$ mm, was 27.3 GHz.

The structure of Figure 2a was used as a reference for the parametric study (parameters: $\varepsilon_r = 3$, $h_{Air} = 0.3$ mm, $h_{Diel} = 1.52$ mm, $p = 2.75$ mm, $A_1 = A_2 = 2.5$ mm, $B_1 = B_2 = 0.75$ mm, $\varphi = \theta = 30°$, and $d = 0$ mm). This section is subdivided into four main subsections, whereby each represents an alternative to break the symmetry and open stop-bands. Rupture of the symmetry is achieved by varying four fundamental parametric relationships: the size ratio between ellipses $A_2/A_1$, the vertical displacement of the second ellipse normalized with respect to the period $d/p$, the orientation of the ellipses $\varphi$ and $\theta$, and the relationship between semi-minor axis $B_1$, $B_2$ and semi-major axis $A_1$, $A_2$ of the ellipses. The graphs also include other parameters that do not play a role in breaking the symmetry, but that can intensify its effect. At the end of this section, the effect of varying the period $p$ is presented, for glide and non-glide configurations. All graphs represent the reference points at the right side of the dispersion diagram ($\beta(2p)/\pi = 1$), as indicated in Figure 2. The continuous lines indicate the lower limit of the stop-band, while the dashed lines indicate the upper limit.

### 2.1. Symmetry Broken by the Ellipse Size

The first observed rupture of the symmetry occurs when the ellipse sizes are different. For simplicity, the relationship between the minor and major semi-axes is maintained constant ($B_1/A_1 = B_2/A_2 = 0.3$). In Figure 3, the stop-band opens when the second ellipse reduces its size, $A_2$, to near zero while keeping the size $A_1$ constant ($A_2/A_1 = 0.1$). It should be noted that the stop-band width begins to reach saturation for values less than $A_2/A_1 = 0.5$. This means that it is not necessary to reach very small ellipse sizes to achieve a larger stop-band, which facilitates the manufacturing. As expected, for an $A_2/A_1 = 1$ ratio, the glide symmetry is recovered and the stop-band is completely closed. Additionally, results are included for different sizes of the first ellipse ($A_1$ from 1.5 mm to 2.5 mm). The value of the period $p$ of the structure was kept constant at 2.75 mm. It can be seen how the reduction in size of this ellipse leads to narrower stop-bands. In addition, the refractive index of the structure increases as $A_1$ increases. This is consistent, as the fields propagating in the air gap are most affected by the dielectric for larger ellipse sizes.

### 2.2. Symmetry Broken by the Displacement between Ellipses

The second rupture of the symmetry relates to the displacement of the second ellipse toward the first ellipse with the period, $d/p$. In combination with this, Figure 4 also shows the effect of varying the thickness of the air gap normalized to period, $h_{Air}/p$. The value of $p$ is 2.75 mm for all the cases. The rupture of the symmetry is greater upon increasing $d/p$ values; thus, the stop-band increases. Of special interest is the significant effect that reducing the thickness of the air gap has on the stop-band widening. The explanation for this phenomenon is once again that the effect of the elliptical structures intensifies when the field is more confined within the air gap. This also explains the shifting to lower frequencies for smaller thicknesses as the refractive index increases.

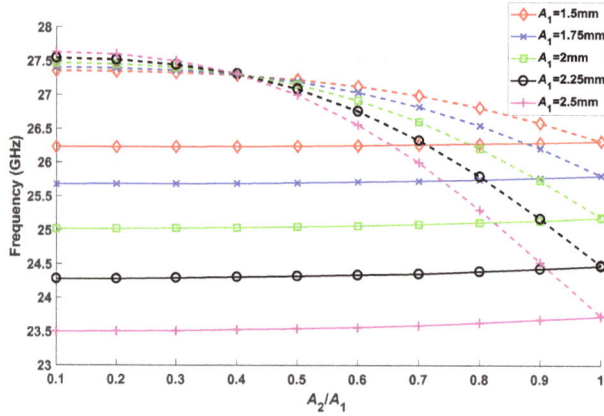

**Figure 3.** Effect of the size of the second ellipse versus the first ellipse $A_2/A_1$ for different values of $A_1$. The rest of the parameters are $\varepsilon_r = 3$, $h_{Air} = 0.3$ mm, $h_{Diel} = 1.52$ mm, $p = 2.75$ mm, $B_1/A_1 = B_2/A_2 = 0.3$, $\varphi = \theta = 30°$, and $d = 0$ mm.

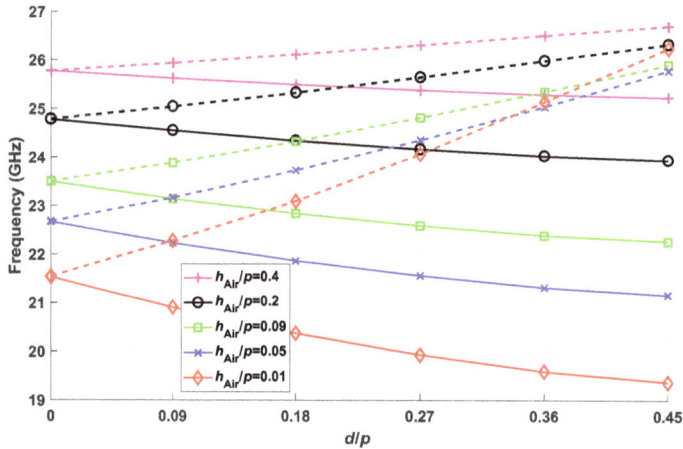

**Figure 4.** Effect of the displacement $d$ of the second ellipse toward the first ellipse for different values of the air gap thickness $h_{Air}/p$. The rest of the parameters are $\varepsilon_r = 3$, $h_{Diel} = 1.52$ mm, $p = 2.75$ mm, $A_1 = A_2 = 2.5$ mm, $B_1 = B_2 = 0.75$ mm, and $\varphi = \theta = 30°$.

### 2.3. Symmetry Broken by the Orientation of the Ellipses

Another alternative to break the symmetry lies in the $\varphi$ and $\theta$ angles at which the ellipses are oriented. Figure 5 depicts the stop-band behavior at a $\theta$ angle from 0° to 90° for $\varphi$ values that also range from 0° to 90° in steps of 10°. In order to have a perspective of the effect of all the parameters of the structure, Figure 5 includes the results for various values of the thickness of the dielectric substrate normalized to the square root of its dielectric constant, $h_{Diel}/\sqrt{\varepsilon_r}$ ($\varepsilon_r = 3$). It can be observed that the curves cross each other when $\varphi$ and $\theta$ are equal, which happens in cases where glide symmetry exists. Therefore, the stop-band opens for $\theta$ values that differ more from $\varphi$. The curves have a sinusoid shape associated with the rotation of the ellipses, for which only a half of the period is represented, since a similar behavior is expected in the other half ($\theta$ values from 90° to 180°). Regarding the thickness of the dielectric substrate, Figure 5 shows how the width of the stop-band increases for greater thicknesses, but this effect tends to saturate from a certain thickness. This is because the mode that penetrates the ellipses into the dielectric is an evanescent mode. Thus, when the thickness is small, the mode does not have enough space to be attenuated; however, for thicknesses greater than a certain threshold, the mode is attenuated enough so that it is not affected by an increase in thickness.

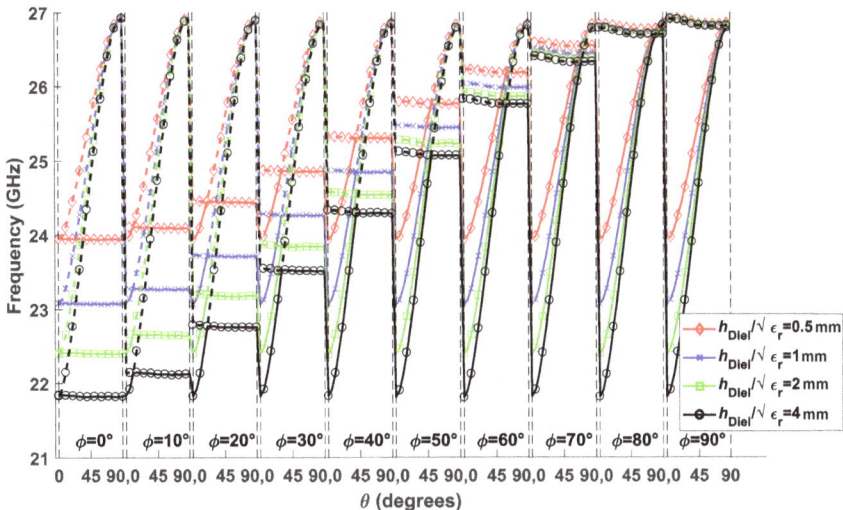

**Figure 5.** Effect of the orientation $\varphi$ and $\theta$ of each of the two ellipses for different values of the dielectric substrate thickness $h_{\text{Diel}}/\sqrt{\varepsilon_r}$. The rest of the parameters are $\varepsilon_r = 3$, $h_{\text{Air}} = 0.3$ mm, $p = 2.75$ mm, $A_1 = A_2 = 2.5$ mm, $B_1 = B_2 = 0.75$ mm, and $d = 0$ mm.

It should be noted that the points corresponding to cases with glide symmetry, where the stop-band closes, rise in frequency from the configuration $\varphi = \theta = 0°$ to the configuration $\varphi = \theta = 90°$. This is equivalent to the reduction of the equivalent refractive index. This is due to the way in which the modes are coupled through the elliptical slots, which depends on their orientation perpendicular or parallel to the direction of propagation. The coupling of a transverse electromagnetic (TEM) mode is minimal in a slot oriented in the direction of propagation ($\varphi = \theta = 90°$), but it is maximum if oriented perpendicularly ($\varphi = \theta = 0°$). Therefore, the effect of dielectric substrate is lower in the first case producing a lower equivalent refractive index.

For clarity, the widest stop-band cases are selected for each $\varphi$ value and represented in Figure 6 as a fractional stop-band. Wider stop-band structures have a $\theta$ value of 0° for $\varphi$ values less than 40°, and 90° for $\varphi$ values greater than 40°. Of all the cases presented, the maximum stop-band is given when $\varphi = 0°$ and $\theta = 90°$ or $\varphi = 90°$ and $\theta = 0°$, which correspond to equivalent structures. In these cases, the width of the stop-band is about twice as wide as for $\varphi = 40°$. In Figure 6, we have a better visualization of the effect of dielectric thickness on the stop-band width. The thickness of the dielectric is doubled in each curve, but the increase in the width of stop-band is progressively reduced.

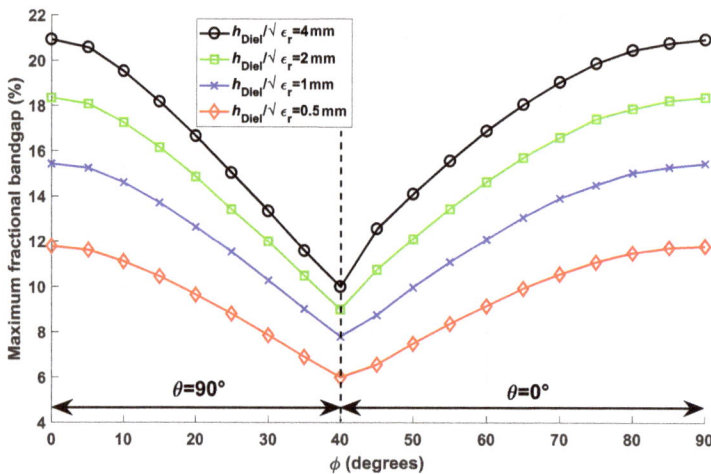

**Figure 6.** Maximum fractional stop-band found for each value of $\varphi$ in Figure 5 considering different substrate thicknesses.

### 2.4. Stop-Band Frequency Shifting and Semi-Minor vs. Semi-Major Axis Relationship

The last parametric study carried out in this work relates to the period of the structure with stop-band frequency displacement. Different values of period $p$ were taken with the corresponding scaling of the ellipse dimensions to preserve the percentage of area occupied by the ellipses in the unit cell. The parameters of the reference structure for this stop-band study are as follows: $h_{Air} = 0.3$ mm, $h_{Diel} = 1.52$ mm, $A_1 = A_2 = 2.5 \cdot (p/2.75)$ mm, and $d = 0$ mm. A parametric sweep of the $B_1$ and $B_2$ values was also included in the study, so that the ratio of the semi-minor and semi-major axes $B/A$ ($B = B_1 = B_2$; $A = A_1 = A_2$) of the ellipses varied from 0.1 to 1. The results are shown in Figure 7a,b. Figure 7a contains the dispersion diagram value at $\beta(2p)/\pi = 1$ of a structure with glide symmetry ($\varphi = 30°$, $\theta = 30°$). The analysis on a broken-symmetry structure ($\varphi = 0°$, $\theta = 90°$) is depicted in Figure 7b.

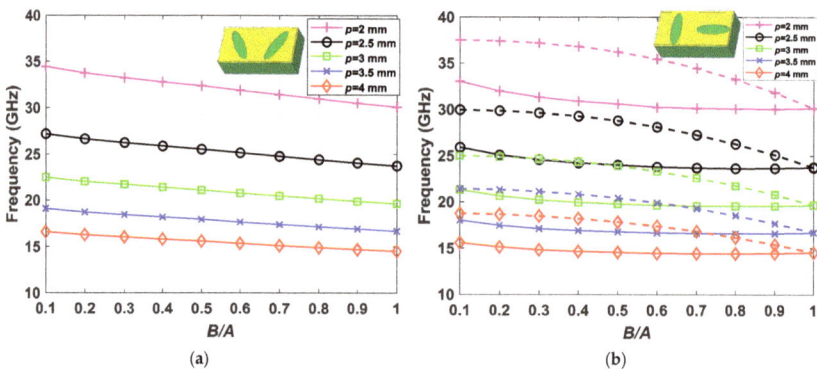

**Figure 7.** Effect of the semi-minor vs. semi-major axis $B/A$ relationship for different values of the period $p$: values for a glide-symmetric structure (**a**) and a non-glide structure (**b**). The rest of the parameters are $h_{Air} = 0.3$ mm, $A_1 = A_2 = 2.5 \cdot (p/2.75)$ mm, and $d = 0$ mm.

It is characteristic of periodic structures that the frequency behavior is inversely proportional to its period. Thus, in Figure 7, it is observed that the modes move downward in frequency with the inverse of $p$. In the glide-symmetric configuration, there is no stop-band; thus, in Figure 7a, only one curve is represented for each periodicity value $p$. It is interesting to note that, when the $B/A$ value is higher, the frequency of the modes decreases, i.e., the refractive index increases. This is because the interaction of the electromagnetic fields with the dielectric material increases as a result of the reduction of the metallic surface. With respect to the non-glide configuration of Figure 7b, both the curve for the starting frequency (solid) and the curve for the ending frequency (dashed) of the stop-band are represented. The pattern is similar to the behavior of the glide configuration with the increase of the $B/A$ value. However, there is a variation in the width of the stop-band that reaches its minimum when $B/A = 1$, in which we recover the symmetry and the stop-band disappears. Figure 8 represents the width of the stop-band obtained in each of the $p$ values represented in Figure 7b. It is clearly observed that the maximum width of the stop-band is achieved for $B/A$ values ranging between 0.3 and 0.4.

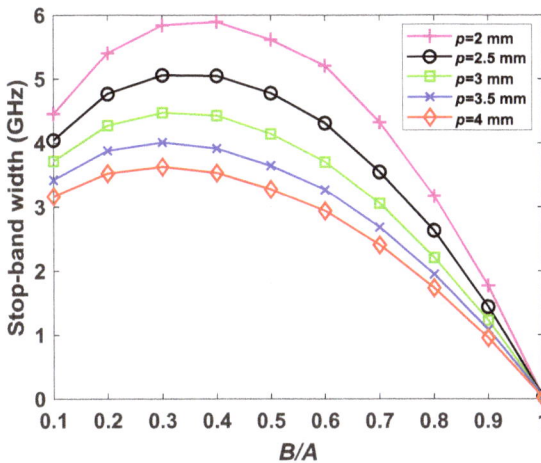

**Figure 8.** Bandwidth of the stop-band for the non-glide configuration in Figure 7b as a function of $B/A$ for different values of periodicity $p$.

### 2.5. Regeneration of the Glide Symmetry

An interesting property of the structures analyzed in this work is the regeneration of glide symmetry from a structure with broken symmetry. The procedure consists of applying the symmetry again on the non-glide unit cell using one of the lateral edges as a symmetry axis. The procedure is shown in Figure 9, where the original unit cell is reflected and translated a distance $p$ in the direction of propagation. Of special interest is the case in which the unit cell with broken symmetry has mirror symmetry, i.e., it is identical to the reflected unit cell considering the center in the $y$-direction as the axis of symmetry. This particularity allows to break and regenerate the symmetry only by displacing a column of the unit cell at periodicity $p$. In periodic prototypes with several unit cells, it implies the capacity to design reconfigurable structures based on the relative displacement of alternated columns in the direction of propagation.

Figure 10 depicts a comparison of the dispersion diagrams obtained with the unit cells of Figure 9a,b. The main stop-band present for $\beta(2p)/\pi = 1$ in the non-glide unit cell, from 20.92 GHz to 26.76 GHz, closes completely by regenerating the symmetry. Another interesting effect on non-glide structures is the appearance of an additional stop-band produced by the merging of the forward mode in air with the reverse mode in the dielectric. In the example shown in Figure 10, this occurs for values around $\beta(2p)/\pi = 0.75$. This type of stop-band was explained in Reference [31] as a complex mode that propagates in the structure. This additional stop-band is also completely closed in the glide configuration.

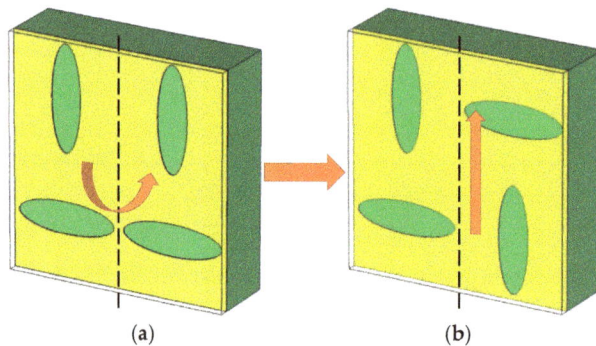

**Figure 9.** The non-glide-symmetric structure in Figure 7b mirrored using a different symmetry axis (**a**); the new glide-symmetric structure (**b**) is obtained from (**a**) after applying a translation.

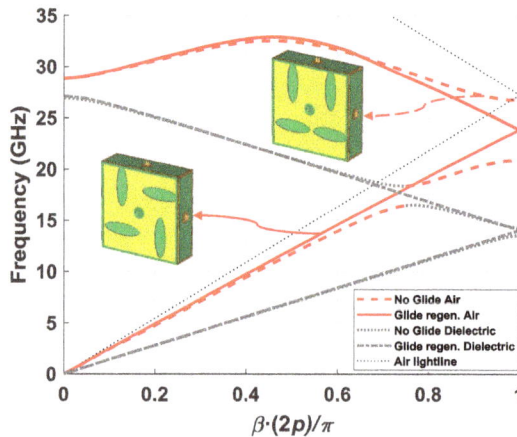

**Figure 10.** Dispersion diagram of the structure in Figure 9b with regenerated symmetry compared to the original non-glide-symmetric structure in Figure 9a.

*2.6. Materials and Instrumentation*

In order to carry out an experimental demonstration of the phenomena described in this section, different devices were used for both the manufacture and the measurement of prototypes. The manufacture of the ellipses on the substrate was carried out with an LPKF ProtoLaser S4 with the dielectric material RO4350B ($\varepsilon_r = 3.66$; tan($\delta$) = 0.0035 @10GHz). The necessary aluminum parts were machined in an external company. Twenty 2.92-mm coaxial connectors with a core diameter of 0.3 mm and length of 7 mm were chosen for the connections. The calibration kit used was the 85056K HP/Agilent. The measurements were made using two 2.92-mm low-phase-error coaxial cables minibend KR-6, and eight loads Anritsu K210. Finally, the vector network analyzer used was the Agilent 8722ES model.

**3. Results**

Four designs were made for the experimental demonstration. In three of them, we used the unit cells shown in Figure 11, which corresponded to the glide configuration (Figure 10), the non-glide configuration (Figure 11b), and the configuration with regenerated glide symmetry from a non-glide configuration (Figure 11c). The parameters chosen for the designs were $\varepsilon_r = 3.66$, $h_{Air} = 0.2$ mm, $h_{Diel} = 1.52$ mm, $p = 2.75$ mm, $A_1 = A_2 = 2.5$ mm, $B_1 = B_2 = 0.75$ mm, and $d = 0$ mm. The only values that were different were $\varphi$ and $\theta$, taking $\varphi = \theta = 30°$ for the glide configuration and $\varphi = 0°$, $\theta = 90°$ for the non-glide and regenerated configurations. The glide (Figure 12a) and non-glide (Figure 12b) designs contained $10 \times 10$ unit cells, whereas the design with regenerated symmetry (Figure 12c) contained only $5 \times 10$ unit cells due to the unit cell being twice as wide. The fourth design (Figure 12d) was a supporting structure that acted as a "thru" and that was used to make phase corrections in the measurements of the other three designs. The prototypes were excited by 10 coaxial connections, five on each side, to create a plane wave that emulated the boundary conditions used in the analyzed unit cells. It should be noted that we added a non-glide frame of ellipses around the four designs to prevent the propagation of spurious modes between the metal surface of the dielectric substrate and the aluminum casing that shielded these designs. As we can observe in Figure 12, the total length of the surface with ellipses was 55 mm (each unit cell was 5.5 mm in length), but the distance between the coaxial connectors and the ellipses was 24 mm in total. This is the separation that should exist between coaxial inputs and outputs of the "thru" for a correct post processing of the measurements.

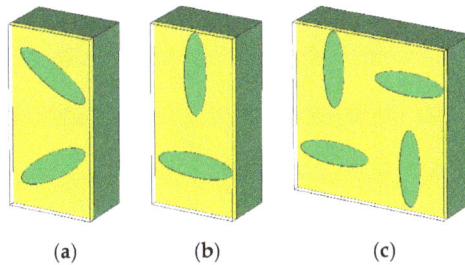

(a)       (b)       (c)

**Figure 11.** Unit cells of the manufactured designs: glide-symmetric configuration (**a**), non-glide configuration (**b**), and regenerated glide configuration (**c**).

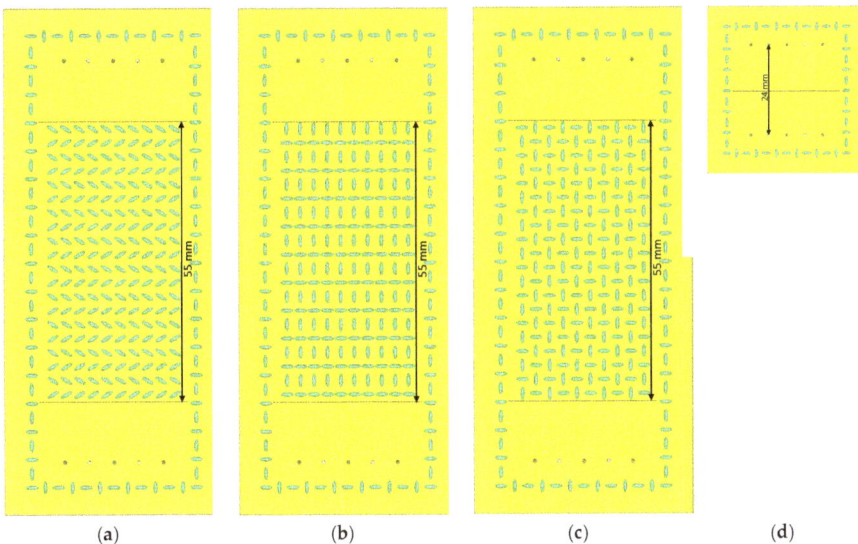

(a)       (b)       (c)       (d)

**Figure 12.** Elliptic configurations of the final designs: glide configuration (**a**), non-glide configuration (**b**), regenerated glide configuration (**c**), and "thru" section (**d**).

Two metal casings, one top and one bottom, covered the dielectric sheet. Ten coaxial connectors were inserted in these casings, six in the top casing and four in the bottom casing. In this way, the excitation at the input through five coaxial probes was achieved by alternating the up–down orientation of the connectors, as can be seen in the Figure 13. By introducing three probes from the top and two from the bottom, it is possible to achieve a separation between contiguous probes of 5 mm, which coincided with half a wavelength in free space at 30 GHz. This ensured a good plane wave across the desired frequency range, between 20 and 30 GHz. Since the base size of the coaxial connectors was 9.5 mm, it was impossible to achieve this separation by inserting the probes on one side only. Therefore, the purpose of these metal casings was threefold. First of all, the 2.92-mm coaxial connectors could be screwed on. Secondly, the internal shape of these casings provided an acceptable matching of the complete prototype between 20 and 30 GHz, as the phenomenon to be demonstrated occurred between these frequencies. Finally, the thickness of each casing (top and bottom) was such that the phase difference between the external surface and the inner surface was the same to ensure that all probes excited the structure in phase. For this, both the thickness of the dielectric and the metallization of both sides of the dielectric were taken into account. All dimensions of the metal casings are shown in detail in Figure 13. The internal structure of the top casing is shown in Figure 13a. This part corresponds to the design of the "thru"; thus, its length was smaller than for the other designs. In particular, this casing increased its length by an additional 55 mm while keeping the 0.2-mm gap in the center of the piece. There was a staggered shape from the input and output with the five holes for the coaxial probes to the central area with the gap of 0.2 mm. In the design, a 3-mm corrugation was introduced along the entire edge, which corresponded to approximately a quarter wavelength at the central frequency. This allowed the short to be transformed into open and again into short-circuit in the internal boundary of the casing, ensuring good electrical isolation of the fields. Figure 13b represents a lateral cut of the complete "thru" design where the coaxial probes entered from the bottom side. Similarly, Figure 13c shows another lateral cut in which the coaxial probes penetrated from the top side. All holes in the dielectric substrate with a diameter of 0.7 mm were metallized to preserve the coaxial structure when the probes were introduced through them. For a better visualization, Figure 13d includes a detailed image of the central zone of Figure 13b,c. Finally, Figure 13e,f present the details of the inner sides of the top and bottom part of the casing. All dimensional values required in the design of the casing are given in Figure 13b–f.

The images of the manufactured prototype are shown in Figure 14a–d. To reduce manufacturing costs, the four designs of Figure 12a–d were implemented on a single piece of dielectric substrate (Figure 14a). In addition, instead of manufacturing a complete metal casing covering the entire prototype in Figure 13a, a single casing was built with the short section for the thru and a long section for the glide (Figure 12a), non-glide (Figure 12b), and regenerated glide (Figure 12c) designs. This allowed a considerable reduction of the manufacturing cost, as the long section of the metal casing was identical in all three cases. This reduced casing consisted of the top and bottom parts, as shown in Figure 14b. Figure 14c shows a detailed photo of the stepped transition and Figure 14d depicts the final assembly in the measurement process with the cables and loads connected. For the measurement of the long designs, we moved the casing to the position of each one. Detailed images of the three ellipse configurations can be found in Figure 15: glide (Figure 15a), non-glide (Figure 15b), and regenerated glide (Figure 15c).

The measurement process was critical, requiring at least 35 measurements for each of the designs. Therefore, all connections had to be very stable to ensure that the conditions were the same in all 35 measurements. The results obtained from these 35 measurements were properly processed to combine the individual S parameters of each port in reflection and transmission parameters of an equivalent simultaneous excitation at the five input ports. This process was applied to the four designs, and the results obtained are represented in Figure 16, together with the results of the simulations. Figure 16a represents the results of the "thru" section, Figure 16b represents the results of the glide configuration section, Figure 16c represents the results of the non-glide section, and Figure 16d represents the results of the regenerated glide-symmetric design. The simulations were carried out including the losses in the dielectric (tan($\delta$) = 0.0035) and in the conductors (copper and aluminum). The results obtained fit very well with the simulations, with the exception of a small upward shift in frequency of approximately 0.5 GHz for all cases. The appearance of a band with an attenuation greater than 20 dB between 22.5 and 27.5 GHz was observed in the non-glide prototype (Figure 16c) with respect to the glide prototype (Figure 16b), whose effect was eliminated by regenerating the symmetry (Figure 16d). As indicated in Figure 10, an additional stop-band can be found at frequencies around 17.5 GHz in the non-glide case. The effect of this secondary stop-band was lower with an attenuation level between 15 and 20 dB, but it was present in the measurement due to the merging of the forward mode in the air and the backward mode in the dielectric. Finally, the appearance of a third stop-band in the three cases studied around 35 GHz was also of interest, which was produced again by the fusion of forward and backward modes in air and dielectric.

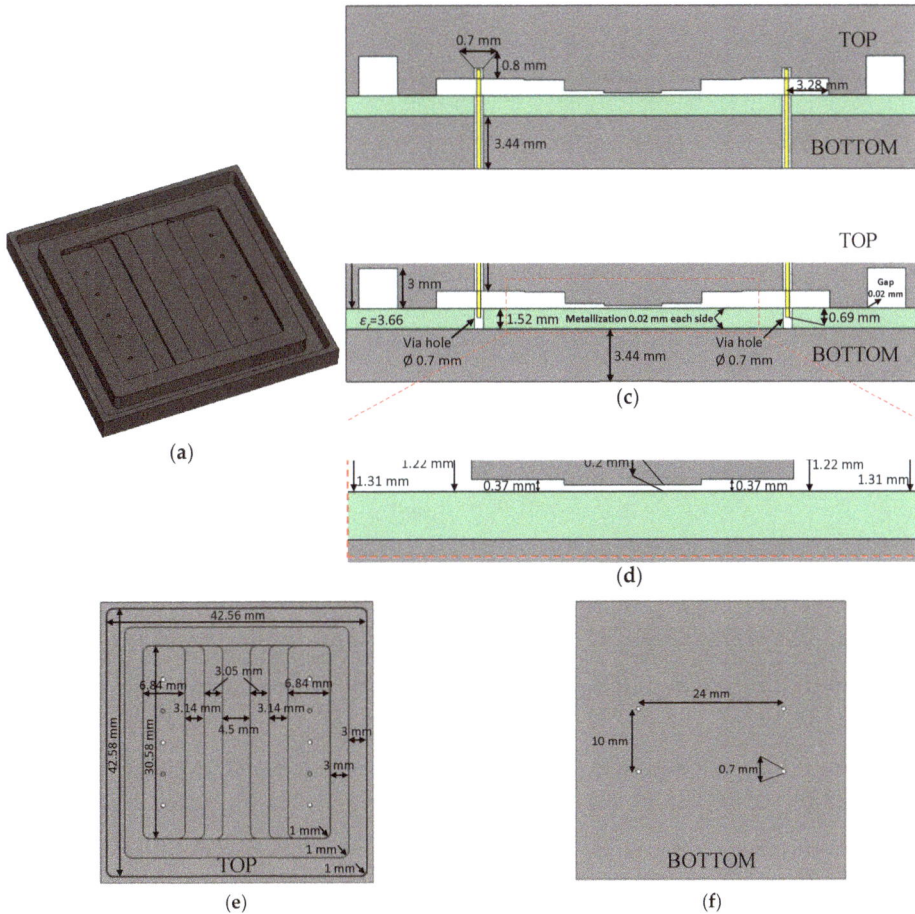

**Figure 13.** Pieces of the metal casing for the "thru" section. Inner view of the top piece (**a**); cross-section with the probes entering from the bottom part (**b**); cross-section with the probes entering from the top part (**c**); detailed view of the stepped transition (**d**); details of the top piece (**e**); details of bottom pieces (**f**).

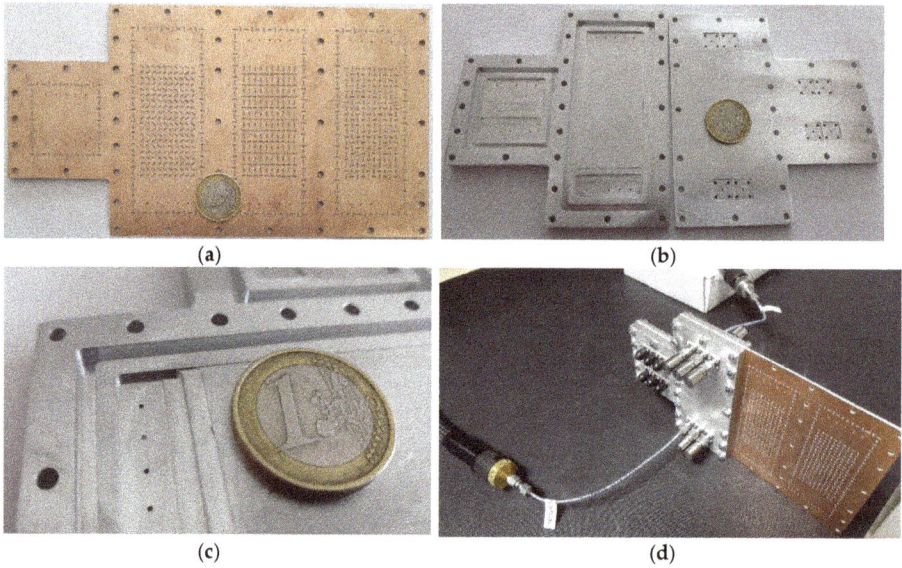

**Figure 14.** Pictures of the manufactured prototypes and assembly. Dielectric substrate with the elliptic patterns (**a**); top and bottom parts of the casing (**b**); detail of the transition at the top part of the casing (**c**); final measurement assembly (**d**).

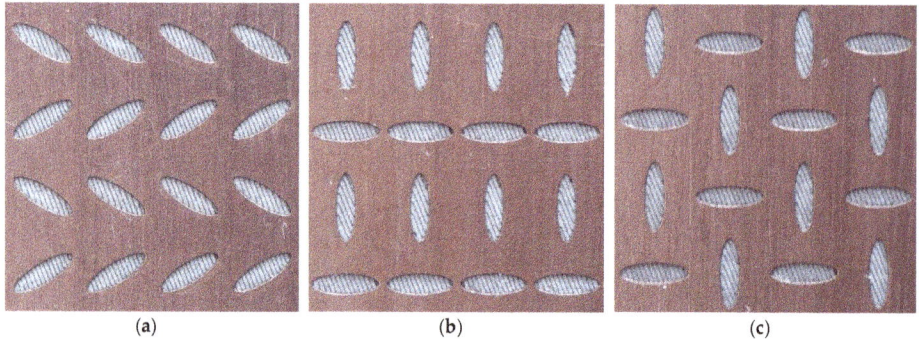

**Figure 15.** Detailed pictures of the elliptic configurations: glide configuration (**a**), non-glide configuration (**b**), and regenerated glide configuration (**c**).

Of equal interest was the effect of these periodic elliptical structures on the phase. These results are depicted in Figure 17. For this study, the phase difference introduced by the "thru" section, represented in Figure 17a, was used as a reference. The first step was to eliminate the phase shift introduced by the "thru" section from the measured phase of the elliptical configurations. With this, we obtained the phase difference produced only by the 55-mm section with ellipses (10 unit cells). These results are represented in Figure 17b–d, applying a normalization in the phase with the factor $\beta(2p)/\pi$. As with the magnitude values, the measured phase values represented in the dispersion diagrams in Figure 17b,c coincided very well with the dispersion diagrams obtained from the unit-cell analysis. The 0.5-GHz upward deviation in frequency appeared again, but the shape exactly fit the results obtained in simulation. In addition to the measurement result, the graphs also included the result obtained from the full wave simulation of the S parameters of the complete designs, which matched better than the dispersion diagrams of the unit cells. In the glide (Figure 17b) and regenerated glide (Figure 17d) cases, a non-dispersive behavior up to 30 GHz was observed. The measured curve was closer to the light line in the regenerated glide configuration than in the original glide configuration, reaching the normalized value of 1 at 24.4 GHz in the first case and at 23.6 GHz in the second case (light line at 27.3 GHz). This means that there was a higher refractive index in the original configuration, which could be modified by varying the parameters of the unit cell. Of special interest was the dispersion diagram measured for the non-glide prototype (Figure 17c), in which the two stop-bands produced around 17.5 GHz and from 22 GHz to 27 GHz were clearly observed. In the range between these two stop-bands (18 to 22 GHz), there was a considerably higher refraction coefficient than in the glide configurations, which could be used to implement a structure with a higher phase-shifting capacity.

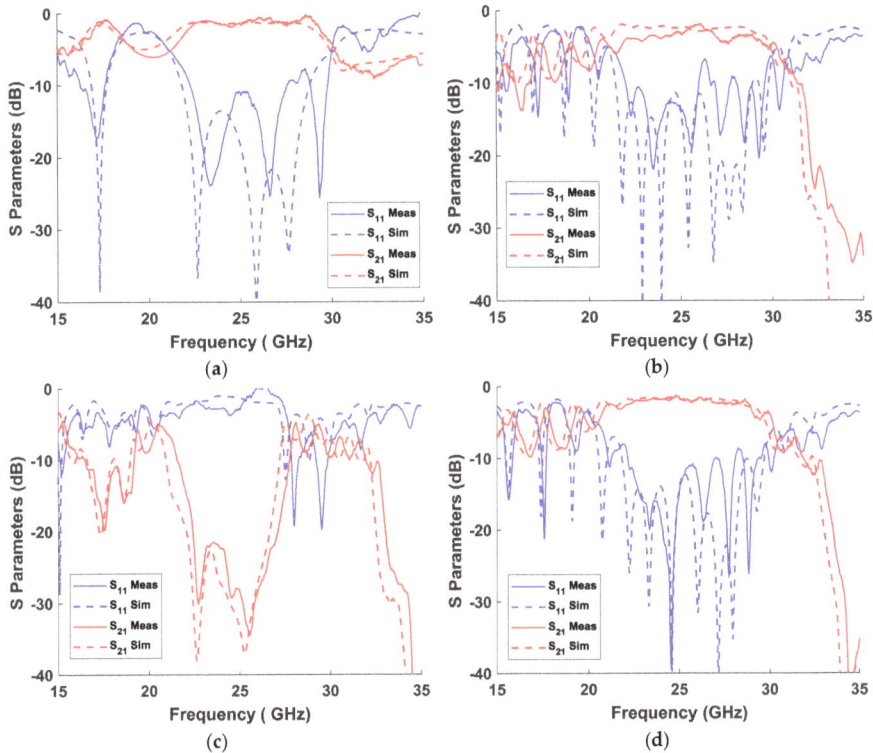

**Figure 16.** Simulated and measured S parameters of the "thru" section (**a**), the glide configuration (**b**), the non-glide configuration (**c**), and the regenerated glide configuration (**d**).

Figure 18 shows a comparison of the behavior of the electric field within the air gap, more precisely in a cut in the middle of the air gap (0.1 mm). The frequencies were conveniently selected to show the most interesting phenomena found in these structures: propagation at frequencies below the appearance of stop-bands (15 GHz), the maximum attenuation of the field found in the secondary stop-band (17.36 GHz), propagation between the two stop-bands (20.22 GHz), the maximum attenuation in the fundamental stop-band (25.28 GHz), propagation at a frequency above the two stop bands (28.86 GHz), and attenuation in the third stop band (35 GHz). The fields propagated homogeneously in the two glide configurations except in the capture at 35 GHz where a stop-band was present in all cases. For the non-glide prototype, the level of the electric field progressively decayed along the structure for the frequencies contained in the stop-bands, especially in the fundamental stop-band (25.28 GHz), which was exhaustively studied throughout this paper.

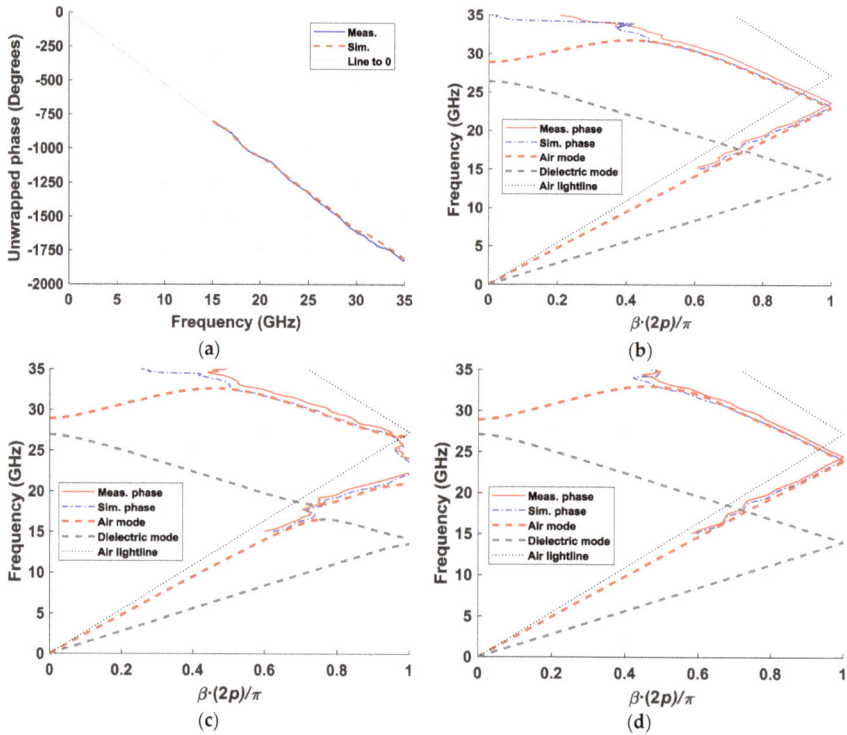

**Figure 17.** Simulated and measured phases in transmission. Phase introduced by the "thru" section (**a**), and dispersion diagrams of the glide (**b**), non-glide (**c**), and regenerated glide (**d**) configurations.

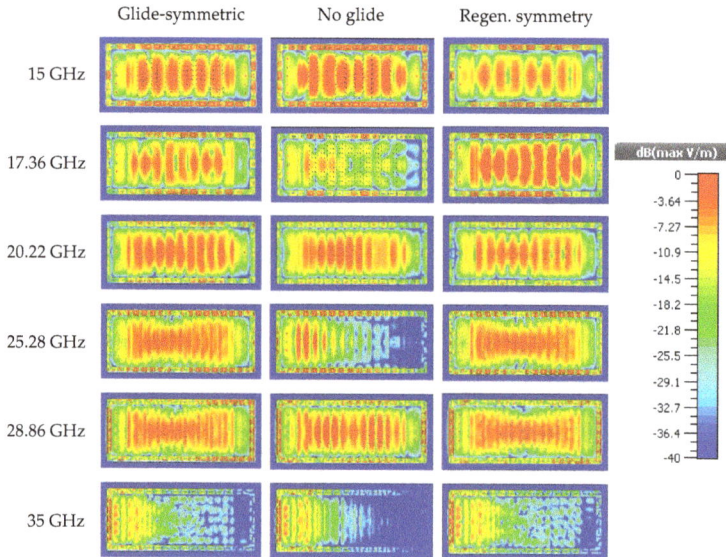

**Figure 18.** Representations of the electric field inside the air gap of the simulated prototypes for the three configurations under study at given frequencies of interest.

## 4. Discussion

This work introduced a new type of glide symmetry based on a single-plane metal structure. The usefulness of these structures lies in generating stop-bands with different levels of attenuation and a variety of refraction coefficients in a simple and easy implementation. This functionality was demonstrated with the analysis of dispersion diagrams and simulation and measurement of the S parameters of the manufactured prototypes. Our parametric study highlights the tremendous number of different configurations that this technology offers to produce stop-bands with different bandwidths at various frequencies, as well as to generate equivalent refractive indexes. Our experimental results showed a stop-band with a bandwidth of 5 GHz centered at the frequency of 25 GHz (20% fractional bandwidth) with an attenuation of 5.5 dB/cm for the non-glide configuration. In the measured glide and regenerated glide cases, a wideband propagation with refractive indexes of 1.12 and 1.16 was obtained, with a possibility to reach values higher than 1.3 with other configurations. The implementation was done using ellipses with a certain orientation on the metallic face of a dielectric substrate. On them, a smooth metal plate was placed at a given distance, providing a gap of air through which the propagation of the electromagnetic fields took place. Unlike previous glide-symmetric structures described in the literature, the alignment of the surface with ellipses and the upper metal plate is not a problem when manufacturing a prototype based on this technology, since a misalignment does not produce any rupture of the symmetry in this case. This is of particular interest for the design and manufacture of Luneburg lenses like the one developed in Reference [26], as it would reduce the manufacturing complexity. The Luneburg lens presented in Reference [26] was implemented by means of two opposing metal parts in which a glide-symmetric holey structure was machined. If the alignment of these two plates is not good enough, ruptures of symmetry may appear between the upper and lower plates, resulting in small stop-bands through the structure. The application of glide symmetry in a single plane prevents the occurrence of this problem. Although we demonstrated in this work that the rupture of the symmetry in the plane of the ellipses generates a stop-band with an attenuation of approximately 30 dB in 55 mm of surface, there are works that demonstrate that the glide-symmetric holey structures in two metallic surfaces provide a greater attenuation per unit length. In particular, the EGB achieved in Reference [23] provided 3.5 times more propagation blockage. This work only studied the attenuation in a particular configuration of ellipses to demonstrate the appearance of stop-bands; thus, further investigation with other configurations is required. The main function of the dielectric material used in the designs of this work is to act as a support for the glide-symmetric ellipses. Future lines of research focused on this type of structures can be based on the substitution of the dielectric substrate by a thin sheet of metal suspended in air with the pattern of ellipses. Since it was demonstrated that it is possible to regenerate glide symmetry from a specific non-glide configuration, the use of thin metal plates instead of dielectric material allows the possibility of breaking and regenerating symmetry by displacing alternate rows of ellipses at a distance equal to the periodicity $p$ (see Figure 10). Another possibility is to replace the dielectric material with an anisotropic material such as liquid crystal [28], whose effective dielectric constant in the direction of propagation can be modified by applying a voltage between the parallel metal plates containing it [32]. This property would make it possible to electrically control the appearance of stop-bands and refractive index. Therefore, these structures offer the possibility of developing both mechanically and electrically reconfigurable devices.

**Author Contributions:** Conceptualization, O.Q.-T.; methodology, A.T.-D.; software, A.T.-D.; validation, A.T.-D.; formal analysis, A.T.-D.; investigation, A.T.-D.; resources, J.-M.F.-G.; data curation, A.T.-D.; writing—original draft preparation, A.T.-D.; writing—review and editing, A.T.-D.; visualization, A.T.-D.; supervision, J.-M.F.-G. and O.Q.-T.; project administration, J.-M.F.-G.; funding acquisition, J.-M.F.-G.

**Funding:** The authors would like to acknowledge the Spanish Government, Ministry of Economy, National Program of Research, Development, and Innovation for the support of this publication in the projects ENABLING-5G "Enabling innovative radio technologies for 5G networks" (project number TEC2014-55735-C3-1-R), FUTURE-RADIO "Radio systems and technologies for high capacity terrestrial and satellite communications in an hyperconnected world" (project number TEC2017-85529-C3-1-R), and the Madrid Regional Government under the project SPADERADAR "Space Debris Radar" (S2013/ICE-3000).

**Acknowledgments:** Simulations done in this work were performed using CST Microwave Studio Suite 2018 under a cooperation agreement with Computer Simulation Technology (CST).

**Conflicts of Interest:** The authors declare no conflict of interest. The funders had no role in the design of the study; in the collection, analyses, or interpretation of data; in the writing of the manuscript, or in the decision to publish the results.

## References

1. Dahlberg, O.; Mitchell-Thomas, R.C.; Quevedo-Teruel, O. Reducing the dispersion of periodic structures with twist and polar glide symmetries. *Sci. Rep.* **2017**, *7*, 10136. [CrossRef] [PubMed]
2. Ghasemifard, F.; Norgren, M.; Quevedo-Teruel, O. Twist and polar glide symmetries: An additional degree of freedom to control the propagation characteristics of periodic structures. *Sci. Rep.* **2018**, *8*, 11266. [CrossRef]
3. Chen, Q.; Ghasemifard, F.; Valerio, G.; Quevedo-Teruel, O. Modeling and dispersion analysis of coaxial lines with higher symmetries. *IEEE Trans. Microw. Theory Tech.* **2018**, *66*, 4338–4345. [CrossRef]
4. Quevedo-Teruel, O.; Dahlberg, O.; Valerio, G. Propagation in waveguides with transversal twist-symmetric holey metallic plates. *IEEE Microw. Wirel. Compon. Lett.* **2018**, *28*, 858–860. [CrossRef]
5. Palomares-Caballero, A.; Padilla, P.; Alex-Amor, A.; Valenzuela-Valdes, J.; Quevedo-Teruel, O. Twist and glide symmetries for helix antenna design and miniaturization. *Symmetry* **2019**, *11*, 349. [CrossRef]
6. Crepeau, P.J.; McIsaac, P.R. Consequences of symmetry in periodic structures. *Proc. IEEE* **1964**, *52*, 33–43. [CrossRef]
7. Hessel, A.; Chen, M.H.R.; Li, C.M.; Oliner, A.A. Propagation in periodically loaded waveguides with higher symmetries. *Proc. IEEE* **1973**, *61*, 183–195. [CrossRef]
8. Kieburtz, R.; Impagliazzo, J. Multimode propagation on radiating traveling-wave structures with glide-symmetric excitation. *IEEE Trans. Antennas Propag.* **1970**, *18*, 3–7. [CrossRef]
9. Quevedo-Teruel, O.; Ebrahimpouri, M.; Kehn, M.N.M. Ultrawideband Metasurface lenses based on off-shifted opposite layers. *IEEE Antennas Wirel. Propag. Lett.* **2016**, *15*, 484–487. [CrossRef]
10. Ebrahimpouri, M.; Quevedo-Teruel, O.; Rajo-Iglesias, E. Design guidelines for gap waveguide technology based on glide-symmetric holey structures. *IEEE Microw. Wirel. Compon. Lett.* **2017**, *27*, 542–544. [CrossRef]
11. Valerio, G.; Sipus, Z.; Grbic, A.; Quevedo-Teruel, O. Accurate equivalent-circuit descriptions of thin glide-symmetric corrugated metasurfaces. *IEEE Trans. Antennas Propag.* **2017**, *65*, 2695–2700. [CrossRef]
12. Ghasemifard, F.; Norgren, M.; Quevedo-Teruel, O. Dispersion analysis of 2-D glide-symmetric corrugated metasurfaces using mode-matching technique. *IEEE Microw. Wirel. Compon. Lett.* **2018**, *28*, 1–3. [CrossRef]
13. Valerio, G.; Ghasemifard, F.; Sipus, Z.; Quevedo-Teruel, O. Glide-symmetric all-metal holey metasurfaces for low-dispersive artificial materials: Modeling and properties. *IEEE Trans. Microw. Theory Tech.* **2018**, *66*, 3210–3223. [CrossRef]
14. Ghasemifard, F.; Norgren, M.; Quevedo-Teruel, O.; Valerio, G. Analyzing glide-symmetric holey metasurfaces using a generalized floquet theorem. *IEEE Access* **2018**, *6*, 71743–71750. [CrossRef]
15. Camacho, M.; Mitchell-Thomas, R.C.; Hibbins, A.P.; Sambles, J.R.; Quevedo-Teruel, O. Designer surface plasmon dispersion on a one-dimensional periodic slot metasurface with glide symmetry. *Opt. Lett.* **2017**, *42*, 3375–3378. [CrossRef]
16. Camacho, M.; Mitchell-Thomas, R.C.; Hibbins, A.P.; Sambles, J.R.; Quevedo-Teruel, O. Mimicking glide symmetry dispersion with coupled slot metasurfaces. *Appl. Phys. Lett.* **2017**, *111*, 121603. [CrossRef]
17. Padilla, P.; Herran, L.F.; Tamayo-Dominguez, A.; Valenzuela-Valdes, J.F.; Quevedo-Teruel, O. Glide symmetry to prevent the lowest stopband of printed corrugated transmission lines. *IEEE Microw. Wirel. Compon. Lett.* **2018**, *28*, 750–752. [CrossRef]
18. Quesada, R.; Martín-Cano, D.; García-Vidal, F.J.; Bravo-Abad, J. Deep subwavelength negative-index waveguiding enabled by coupled conformal surface plasmons. *Opt. Lett.* **2014**, *39*, 2990–2993. [CrossRef]
19. Cavallo, D.; Felita, C. Analytical formulas for artificial dielectrics with nonaligned layers. *IEEE Trans. Antennas Propag.* **2017**, *65*, 5303–5311. [CrossRef]
20. Cavallo, D. Dissipation losses in artificial dielectric layers. *IEEE Trans. Antennas Propag.* **2018**, *66*, 7460–7465. [CrossRef]

21. Chang, T.; Kim, J.U.; Kang, S.K.; Kim, H.; Kim, D.K.; Lee, Y.H.; Shin, J. Broadband giant-refractive-index material based on mesoscopic space-fillingcurves. *Nat. Commun.* **2016**, *7*, 12661. [CrossRef] [PubMed]

22. Jia, D.; He, Y.; Ding, N.; Zhou, J.; Du, B.; Zhang, W. Beam-steering flat lens antenna based on multilayer gradient index metamaterials. *IEEE Antennas Wirel. Propag. Lett.* **2018**, *17*, 1510–1514. [CrossRef]

23. Ebrahimpouri, M.; Rajo-Iglesias, M.; Sipus, Z.; Quevedo-Teruel, O. Cost-effective gap waveguide technology based on glide-symmetric holey EBG structures. *IEEE Trans. Microw. Theory Tech.* **2018**, *66*, 927–934. [CrossRef]

24. Ebrahimpouri, M.; Algaba-Brazalez, A.; Manholm, L.; Quevedo-Teruel, O. Using glide-symmetric holes to reduce leakage between waveguide flanges. *IEEE Microw. Wirel. Compon. Lett.* **2018**, *28*, 473–475. [CrossRef]

25. Rajo-Iglesias, E.; Ebrahimpouri, M.; Quevedo-Teruel, O. Wideband phase shifter in groove gap waveguide technology implemented with glide-symmetric holey EBG. *IEEE Microw. Wirel. Compon. Lett.* **2018**, *28*, 476–478. [CrossRef]

26. Quevedo-Teruel, O.; Miao, J.; Mattsson, M.; Algaba-Brazalez, A.; Johansson, M.; Manholm, L. Glide-symmetric fully-metallic Luneburg lens for 5G communications at Ka-band. *IEEE Antennas Wirel. Propag. Lett.* **2018**, *17*, 1588–1592. [CrossRef]

27. Quevedo-Teruel, O.; Ebrahimpouri, M.; Ghasemifard, F. Lens antennas for 5G communications systems. *IEEE Commun. Mag.* **2018**, *56*, 36–41. [CrossRef]

28. Mueller, S.; Penirschke, A.; Damm, C.; Scheele, P.; Wittek, M.; Weil, C.; Jakoby, R. Broad-band microwave characterization of liquid crystals using a temperature-controlled coaxial transmission line. *IEEE Trans. Microw. Theory Tech.* **2005**, *53*, 1937–1945. [CrossRef]

29. Mesa, F.; Rodríguez-Berral, R.; Medina, F. On the computation of the dispersion diagram of symmetric one-dimensionally periodic structures. *Symmetry* **2018**, *10*, 307. [CrossRef]

30. Nosrati, M.; Daneshmand, M. Substrate Integrated Waveguide L-Shaped Iris for realization of transmission zero and evanescent-mode pole. *IEEE Trans. Microw. Theory Tech.* **2017**, *65*, 2310–2320. [CrossRef]

31. Valerio, G.; Galli, A.; Wilton, D.R.; Jackson, D.R. An enhanced integral-equation formulation for accurate analysis of frequency-selective structures. *Int. J. Microw. Wirel. Technol.* **2012**, *4*, 365–372. [CrossRef]

32. Alex-Amor, A.; Tamayo-Domínguez, A.; Palomares-Caballero, A.; Fernández-González, J.M.; Padilla, P.; Valenzuela-Valdés, J.; Palomares, A. Analytical approach of director tilting in nematic liquid crystals for electronically tunable devices. *IEEE Access* **2019**, *7*, 14883–14893. [CrossRef]

*symmetry*

MDPI

*Article*

# Twist and Glide Symmetries for Helix Antenna Design and Miniaturization

Ángel Palomares-Caballero [1,*], Pablo Padilla [2], Antonio Alex-Amor [1], Juan Valenzuela-Valdés [2] and Oscar Quevedo-Teruel [3]

[1] Department of Languages and Computer Science, Universidad de Málaga, 29071 Málaga, Spain; aalex@lcc.uma.es
[2] Department of Signal Theory, Telematics and Communications, Universidad de Granada, 18071 Granada, Spain; pablopadilla@ugr.es (P.P.); juanvalenzuela@ugr.es (J.V.-V.)
[3] Department of Electromagnetic Engineering, KTH Royal Institute of Technology, SE-100 44 Stockholm, Sweden; oscarqt@kth.se
* Correspondence: angelpc@uma.es; Tel.: +34-958248899

Received: 26 January 2019; Accepted: 4 March 2019; Published: 8 March 2019

**Abstract:** Here we propose the use of twist and glide symmetries to increase the equivalent refractive index in a helical guiding structure. Twist- and glide-symmetrical distributions are created with corrugations placed at both sides of a helical strip. Combined twist-and glide-symmetrical helical unit cells are studied in terms of their constituent parameters. The increase of the propagation constant is mainly controlled by the length of the corrugations. In our proposed helix antenna, twist and glide symmetry cells are used to reduce significantly the operational frequency compared with conventional helix antenna. Equivalently, for a given frequency of operation, the dimensions of helix are reduced with the use of higher symmetries. The theoretical results obtained for our proposed helical structure based on higher symmetries show a reduction of 42.2% in the antenna size maintaining a similar antenna performance when compared to conventional helix antennas.

**Keywords:** higher symmetries; periodic structures; glide symmetry; twist symmetry; dispersion diagram; microwave printed circuits; helix antennas

---

## 1. Introduction

Symmetrical structures are present in a huge number of natural phenomena. On occasion, symmetries have a positive impact in the physical response and properties of materials. Therefore, when symmetries are not spontaneously found in nature, engineers have found a manner to tailor them in an artificial manner [1]. The use of symmetrical geometries modifies the physical properties of materials, such as their mechanical, thermal or electromagnetic responses [2]. Among these properties, those related to electromagnetic behaviour are of great importance for conductive materials and dielectric substrates. The electromagnetic properties resulting from configurations with one-dimensional higher symmetries were initially studied in the 1960s and 1970s [3,4]. However, it has been in recent years, with the new developments on computational electromagnetics, that more complex structures, including two-dimensional and three-dimensional higher-symmetric structures, have been studied. These structures have demonstrated new possibilities for the design of microwave and millimeter-wave circuits and antennas [5]. Some of the key advantages of employing higher symmetries in the design of radiofrequency devices are: significant reduction of the frequency dispersion [6], accurate control of the equivalent refractive index [7], control and elimination of stop-bands in periodic structures [8].

A periodic structure possesses a higher symmetry when it is invariant under a translation and another spatial operator such as a rotation or mirroring. Two commonly used higher symmetries are glide and twist symmetries as illustrated in Figure 1. While the extra spatial operator for glide symmetries is a mirroring with respect to a glide line/surface [9,10], for the twist symmetries the operator is a rotation along a twist axis [10–12]. Twist symmetry is a more general concept than glide symmetry. For example, when the unit cell is symmetrical along the transversal direction, a 180° twist-symmetrical structure turns out to be glide-symmetrical too.

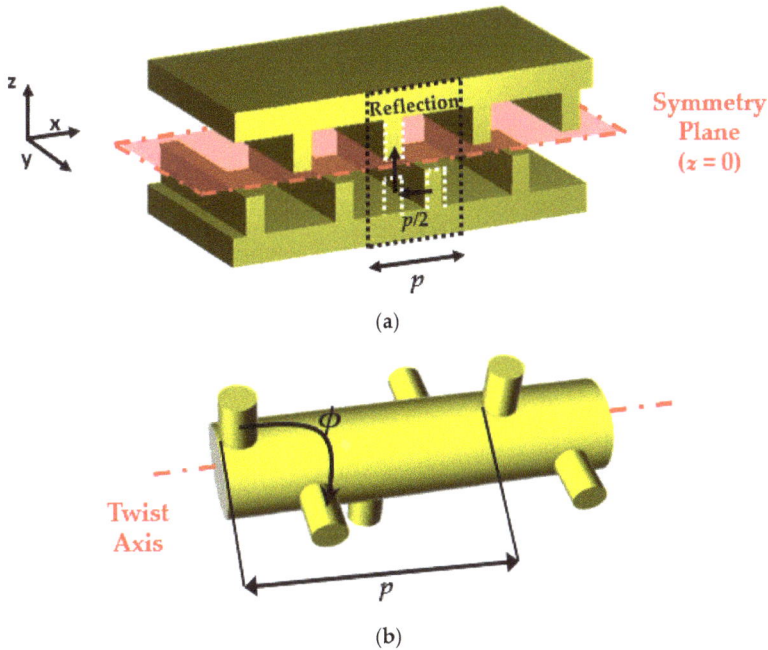

(a)

(b)

**Figure 1.** Examples of higher symmetries: (a) glide-symmetrical corrugations. The corrugations are periodic along *x* axis, and the mirroring plane is *z* = 0; (b) Twist-symmetrical metallic rod with inclusions rotated $\varphi = 90°$ along the twist axis.

Twist and glide symmetries were demonstrated to reduce the dispersion of metasurfaces, providing unit cells with a flattened frequency dependence. This is advantageous for the design of wideband microwave devices, such as planar lenses [3]. Also, the use of glide symmetries in metallic holey structures, as reported in [4], increases the bandwidth of the electromagnetic bandgaps. These glide-symmetrical holey structures have been proposed for cost-effective gap waveguide technology at the millimeter-wave frequency range. Additionally, the combination of twist- and glide-symmetrical configurations increases the equivalent refractive index of periodic structures [5], enhances the linearity of the modes, and creates additional stop-bands at given frequencies [10]. In [13], the use of higher symmetries has been applied to produce a multibeam Luneburg lens antenna with low scan-losses and wide bandwidth. Gap waveguides with glide-symmetrical holey EBG structures were proposed in [14], creating cost-effective guiding structures at high frequencies. Using this technology, a wideband phase shifter was proposed in [15]. Finally, low-loss waveguide flanges were proposed in [16], with the use of glide-symmetric holes around the waveguide apertures. In this manner, the leakage at the waveguide joint is avoided. Lastly, a recent work about the application of glide-symmetry in printed double-sided parallel-strip lines demonstrated the potential of glide symmetry to produce low dispersion transmission lines and filter behavior by breaking the symmetry [9].

Here, we proposed the combination of glide and twist symmetries to reduce the size of helix antennas based on the increment of the equivalent refractive that the higher symmetries produce.

## 2. Materials and Methods

For antenna designs that are based on field guidance and progressive matching towards free space, as is the case in helix-based antennas, the use of higher-symmetrical configurations can be beneficial. In this section, we analyze the electromagnetic effects of including twist and glide corrugations of a metallic flat strip in a helix structure, defining the guidelines for antenna designs.

### 2.1. Baseline Helix Antenna Designs

The general structure of a helix antenna is shown in Figure 2a, and it is formed by a ground plane and a conducting helix structure. The basic parameters that define a helix antenna are: the radius ($r$), the pitch ($p$) and the pitch angle ($\alpha$). The mathematical relation between these parameters is:

$$p = 2\pi r \tan(\alpha) \tag{1}$$

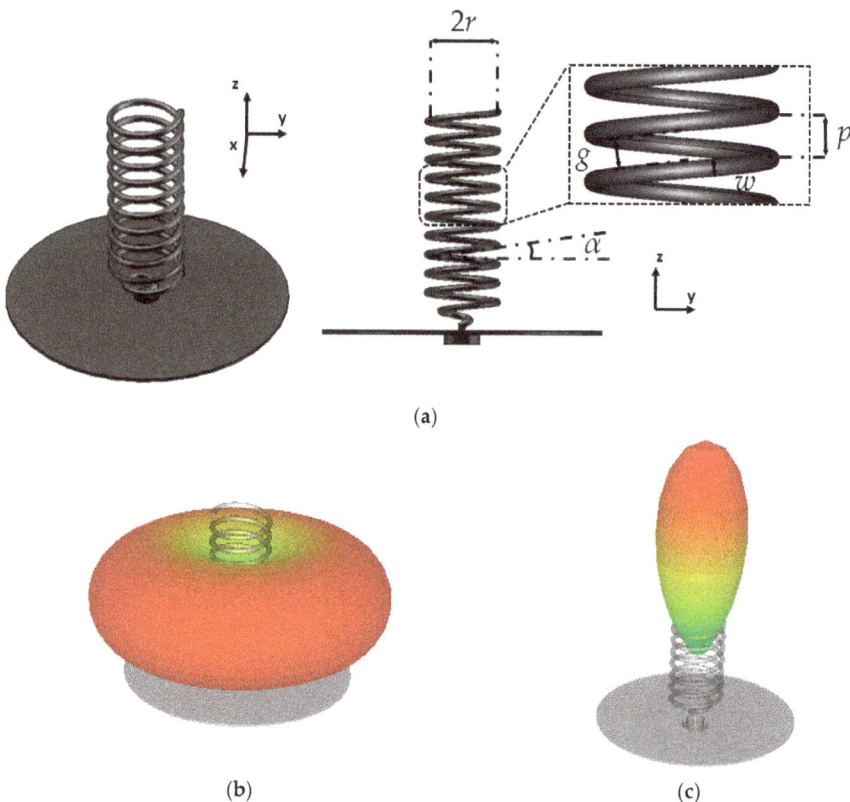

(a)

(b)

(c)

**Figure 2.** Conventional helix antenna composed of a wire and a ground plane: (**a**) antenna model and scheme; (**b**) normal mode of radiation; (**c**) axial mode of radiation.

Depending on these parameters, the antenna can operate in two common modes of radiation: normal and axial mode. The antenna is operating in its normal mode (Figure 2b) if, for a certain working frequency $f_0$, the helix circumference is considerably smaller than the wavelength

($2\pi r << \lambda_0$) [17]. The polarization in this mode is typically aimed to be circular. The antenna operates in its axial mode if the helix circumference is in the order of the wavelength ($2\pi r \approx \lambda_0$) and the pitch distance $p$ is a quarter of the wavelength ($p \approx \lambda_0/4$) [18]. Therefore, for the axial mode, the pitch angle must be around 15 degrees, following Equation (1). This mode produces high directivity as illustrated in Figure 2c and provides circular polarization. The use of twist and glide symmetries, depicted in the following subsections, increase the propagation constant. This can be of interest for antenna miniaturization for a given frequency. The miniaturization of the helix antenna in axial mode is useful in applications with space restrictions, such as in antenna arrays.

A helix antenna can be made with a metallic strip instead of a wire. In this case, the strip width $w$ plays a role on the antenna impedance matching [19]. Equations (2) and (3) show the mathematical relations of the pitch length ($L_p$) and gap between turns ($g$) regarding the basic helix parameters.

$$L_p = \frac{2\pi r}{\cos(\alpha)} \tag{2}$$

$$g = p\cos(\alpha) - w \tag{3}$$

Helix antennas are travelling wave antennas that can be classified as slow-wave since the phase velocity of the wave in the structure is smaller than the speed of light. However, this kind of antennas radiate in the curvatures similar to a conventional helix antenna. This radiation is affected by the discontinuities produced by the corrugations. A classical explanation of the operation of a helix antenna operating in the axial mode is that the radiation is similar to an array of loops whose phase distribution produces an end-fire radiation [20].

## 2.2. Periodic Glide-Symmetrical and Twist-Symmetrical Unitary Cells

The unitary cell corresponds to one turn of the helix. The basic parameters to be considered for the baseline periodic cell are: the periodicity ($p$), the metallic helix strip width ($w$), the gap space between the wrapped helix strip ($g$), and the unwrapped length of one turn of the helix strip ($L_p$). Figure 3a shows the baseline cell both in 3D and in a planar view.

The twist symmetry can be easily introduced to this basic helix cell by adding an integer number of corrugations per cell to any (or both) sides of the strip. These corrugations have inherent twist symmetry due to the helical configuration of the strip. Figure 3b depicts the parameters that characterize the twist-symmetrical strip, which are: the width of the corrugation ($w_c$), its length ($h_c$), and the number of corrugations per unit cell ($N_c$). The unit cell is composed by sub-cells that consist of a pair of adjacent corrugations.

A twist configuration can be modified to become also a glide. In this case, one corrugation of each pair of the consecutive corrugations is moved to the opposite side of the strip, as illustrated in Figure 3c. In that case, the glide-symmetric periodicity includes a pair of opposed corrugations and its unwrapped length is $L_p/N_{subcell}$. Therefore, the number of glide periods is $N_{subcell} = N_c/2$. It should be noted that, although this configuration is glide in its unwrapped version, once it is rolled, the relation between corrugations of the consecutive turns are not glide. This can be corrected by including a small misalignment (offset) of the corrugations in one of the strip sides so the corrugations of the structure, once wrapped, fit together again. The offset value that satisfies this condition is $p \cdot \sin(\alpha) \cdot L_p/N_c$ as illustrated in Figure 3d. Notice that this case keeps a glide configuration, not regarding the strip, but between strip turns.

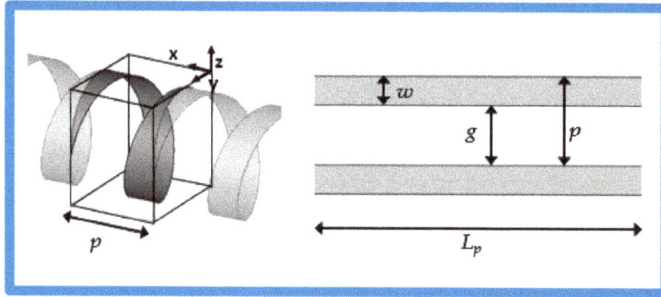

(a)

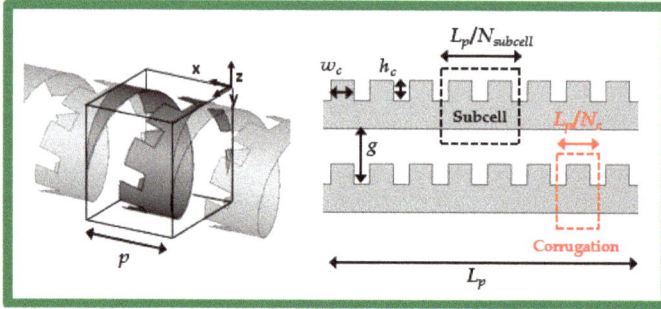

(b)

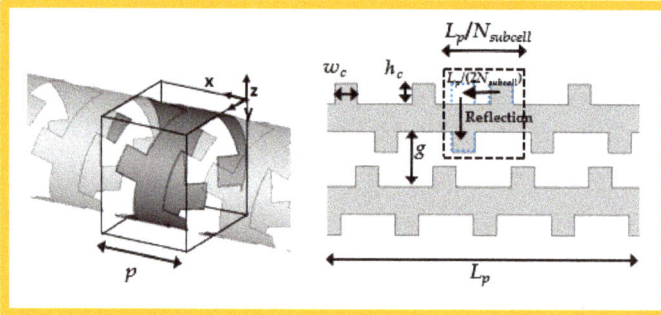

(c)

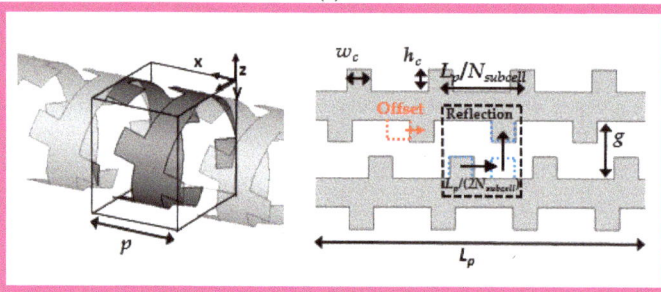

(d)

**Figure 3.** *Cont.*

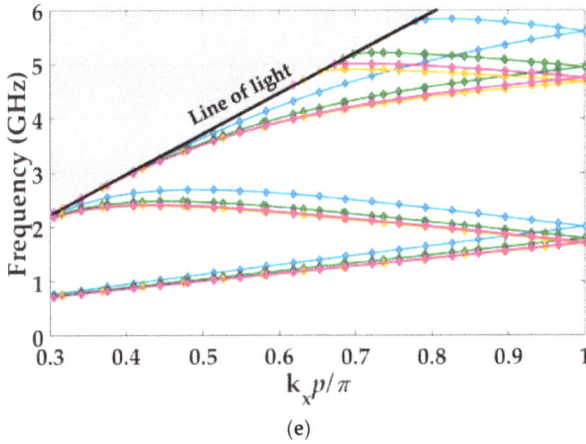

**Figure 3.** Periodic helix cell: (**a**) conventional helix (without corrugations); (**b**) twist-symmetrical helix; (**c**) combined twist- and glide-symmetrical helix; (**d**) combined twist-and-glide-symmetric helix with offset in the glide configuration; (**e**) Their dispersion diagrams. The reference dimensions are: $r = 12$ mm, $\alpha = 15°$, $p = 20.2$ mm, $L_p = 78.06$ mm, $g = 12.78$ mm, $w = p/3$, $N_c = 8$, $h_c = 0.4g$ and $w_c = 0.3L_p/N_c$.

The electromagnetic behavior of this periodic structure can be characterized with the dispersion diagram of the unitary cell that, in this work, is calculated with the eigenmode analysis of CST Microwave Studio. Since this full-wave electromagnetic solver does not allow eigenmode simulations with open boundary conditions, this requirement in the $y$ and $z$ directions is satisfied by oversizing the simulation box in those directions and imposing an electric wall ($E_t = 0$) at these boundaries. This does not have influence in the frequency range under study since the distance to the box boundaries is much larger than the helix diameter. In the $x$ direction (the direction of the axis of the helix structure), the boundary condition is periodic.

Figure 3a–d provide the schematics under study of the helix unitary cell. Their dispersion diagrams are illustrated in Figure 3e. In this graph, the reference is the dispersion diagram of the helix cell without corrugations (blue line). The twist inclusions in the helix unitary cell increase the propagation constant value, which also becomes almost linear with respect to the frequency for the first propagating mode. Additionally, the closed stop-band between the first and second propagation mode is eliminated. The combination of twist and glide symmetries in the unitary cell slightly increases these effects. The glide case with offset presents similar results for the first mode, as illustrated in Figure 3e, but it permits a further increase of the length of the corrugations without the overlapping between turns.

### 2.3. Parametric Tuning Effects

The effect of modifying the parameters of the corrugated structure is here studied. This study is aimed to produce general guidelines that will be of use for the design of helix antennas.

#### 2.3.1. On the Twist Symmetry

The first parametric study, illustrated in Figure 4, is focused on the corrugations of the twist configuration. First, the results for the variation of the length of the corrugations ($h_c$) are depicted in Figure 4a. The increase in the length produces a higher propagation constant, which means an increase of the effective refractive index of the structure. Figure 4b provides the results of modifying the width $w_c$, while preserving the number of corrugations per turn ($N_c$). This effect is smaller than in the case of the length of the corrugations ($h_c$). In Figure 4c, we illustrate the effect of the number of corrugations per turn ($N_c$) for a fixed corrugation width. Although the cell periodicity is the same for all the cases, the different dimensions and number of corrugations introduce a variation in the propagation constant. The parametric study carried out in Figure 4c reveals that the increase in the number of corrugations per turn has a small influence in the propagation constant.

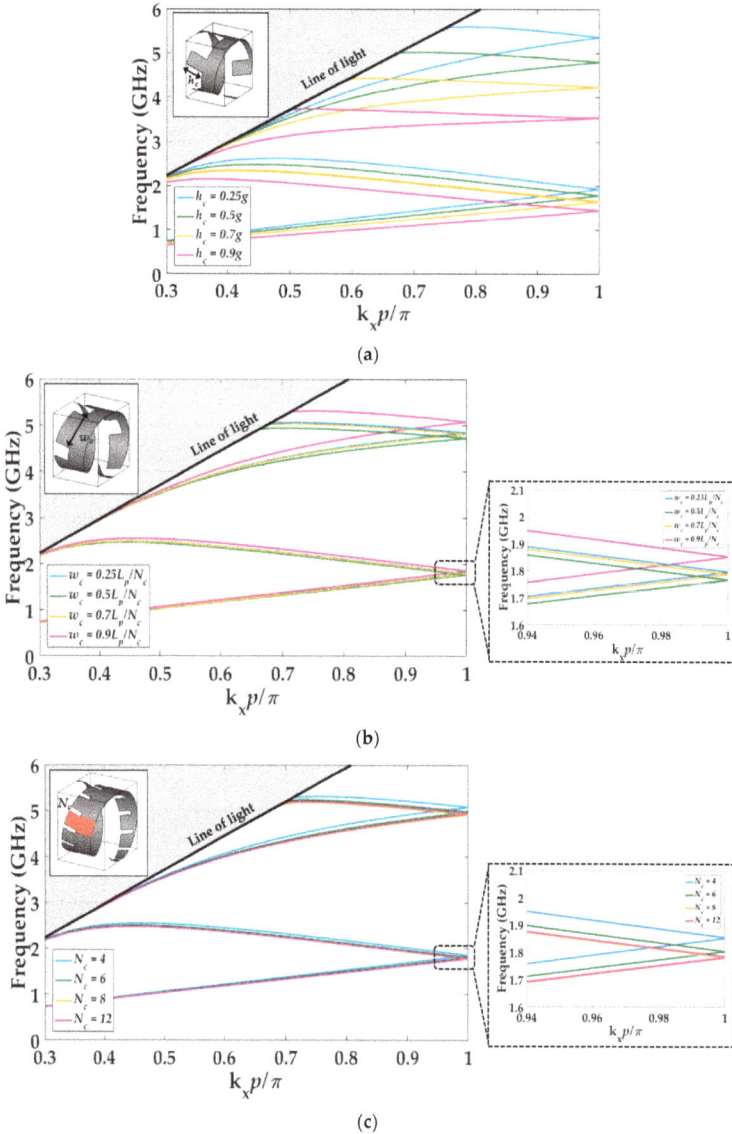

(a)

(b)

(c)

**Figure 4.** Simulated dispersion diagrams for the twist-symmetrical helix cell: (**a**) modification in the length of the corrugations ($h_c$); (**b**) modification of the width of the corrugations ($w_c$), while preserving the number of corrugations per turn ($N_c = 4$); (**c**) modification of the number of corrugations ($N_c$), for a given width, $w_c = 0.06L_p$. The reference dimensions are: $r = 12$ mm, $\alpha = 15°$, $p = 20.2$ mm, $L_p = 78.06$ mm, $g = 12.78$ mm, $w = p/3$, $h_c = 0.4g$ and $w_c = 0.3L_p/N_c$.

### 2.3.2. On the Combined Twist and Glide Symmetry

If glide symmetry is added, the effect previously reported in the twist case is enhanced as depicted in Figure 5a. In this configuration, a higher integration between corrugations is possible, so a higher value for the length of the corrugation can be achieved ($h_c$). Variations in the width of the corrugations (Figure 5b) and the number of corrugations (Figure 5c) have a more limited effect.

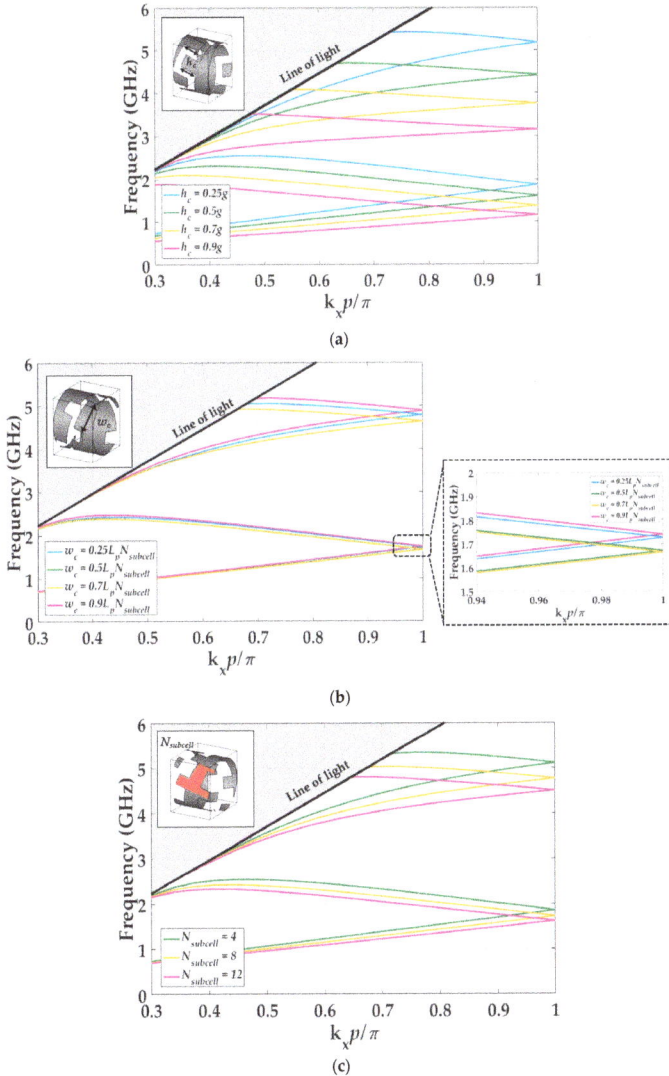

(a)

(b)

(c)

**Figure 5.** Simulated dispersion diagrams for the twist- and glide-symmetrical helix cell: (**a**) modification of the length of the corrugations ($h_c$); (**b**) modification of the width of the corrugations ($w_c$), while preserving the number of glide periods ($N_{subcell}$ = 4) per turn; (**c**) modification of the number of glide periods ($N_{subcell}$), for a given width, $w_c = 0.06L_p$. The reference dimensions are: $r$ = 12 mm, $\alpha$ = 15°, $p$ = 20.2 mm, $L_p$ = 78.06 mm, $g$ = 12.78 mm, $w = p/3$, $h_c$ = 0.4$g$ and $w_c = 0.3L_p/N_c$.

### 2.4. Symmetry Breakage

Finally we illustrate here the effect of the breakage of the higher symmetry in the structure. The presence of the bandgap when the symmetry is broken is useful for filtering purposes [9]. Although this symmetry breakage can be introduced in different ways, here we only show the effect of one representative case that is depicted in Figure 6a. The breakage is produced by increasing the length of the corrugation of one sub-cell. The dispersion diagram of this structure is illustrated in Figure 6b, in which a stop-band between the first and second modes is generated due to the rupture of the symmetry. Additionally, when the symmetry is broken, the frequency linearity of the propagation constant of the first mode is lost. Notice that in this subsection we only present a brief description of the possibility of breaking the symmetry for filtering purposes, but it is not used in the antenna design. In our specific antenna design, the main goal is miniaturization and filtering is not intended. Therefore, the importance relies in the position of the modes in the dispersion diagram that has to be in a lower position regarding to the position of the modes of the reference helix cell. Including a twist-and-broken glide-symmetrical unit cell in the helix antenna would imply a reduction of the operational bandwidth. Deeper studies regarding symmetry breakage can be found in [9].

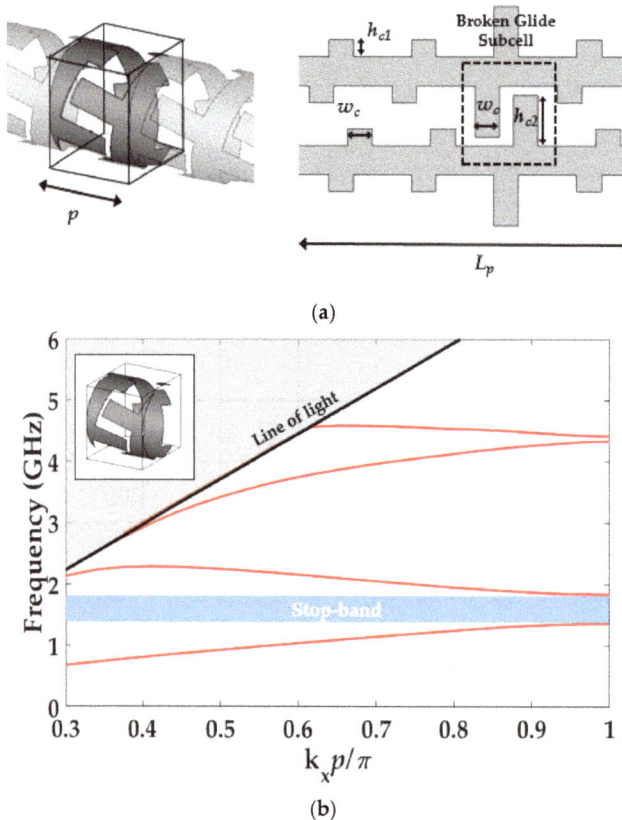

(a)

(b)

**Figure 6.** Symmetry breakage: (**a**) twist-and-broken glide-symmetrical unit cell; (**b**) dispersion diagram. The reference dimensions are: $r = 12$ mm, $\alpha = 15°$, $p = 20.2$ mm, $L_p = 78.06$ mm, $g = 12.78$ mm, $w = p/3$, $N_c = 8$, $h_{c1} = 0.3g$, $h_{c2} = 0.9g$ and $w_c = 0.3L_p/N_c$.

## 3. Helix Antenna Miniaturization

The miniaturization of helix antennas has been the focus of several research studies, mainly for quadrifilar helix antennas [21–24]. The techniques that have been generally employed for their miniaturization are core dielectric loadings [25,26], meander lines [27] and loading of the pointed ends of the helix arms [28].

A straightforward consequence of increasing the propagation constant of the unitary helix cell is the miniaturization of the resulting helix antenna. A miniaturized helix antenna formed by twist-and-glide unitary cells is expected to radiate similar to a strip-only helix antenna, which is larger in dimension.

The following subsections show the comparison between two helix antenna designs operating at the same frequency. The first design is a conventional axial-mode helix strip antenna and the second design is the equivalent miniaturized twist-and-glide symmetric axial-mode helix antenna. Our proposed technique for miniaturization is fully-metallic, which means that there is no dielectric material in the structure. Therefore, dielectric losses are avoided and our miniaturized antenna can reach higher efficiency and gain at high frequency when compared to miniaturized antennas based on dielectric solutions.

### 3.1. Conventional Helix Antenna Design

A conventional helix antenna made of strips can be designed based on the principles described in Section 2.1. In this case, the conventional strip helix antenna design has been designed to operate in 2.45 GHz. The radius of the helix is fixed at $r$ = 19.5 mm, which is the parameter that mainly defines the antenna frequency. The pitch angle $\alpha$ has a value of 15 degrees to obtain a radiation in axial mode. The strip width ($w$) has a value of $p/3$. This parameter has no noticeable effect in the working frequency. The number of helix turns $N_{turns}$ is chosen to be 12. A ground plane is located at the bottom of the antenna with a side dimension around one wavelength at the working frequency. A coaxial cable and a standard SMA-type transition, both of 50 $\Omega$, are considered for the antenna feeding. These dimension values for the unit cell imply that the third mode is the one excited in our conventional antenna design.

Figure 7a illustrates the $|S_{11}|$ calculated in CST Microwave Studio. The circular polarization bandwidth is illustrated in grey. The circular polarization bandwidth represents the frequency range where the antenna has an axial ratio (AR) below 3 dB at the main direction of radiation and operates in axial mode.

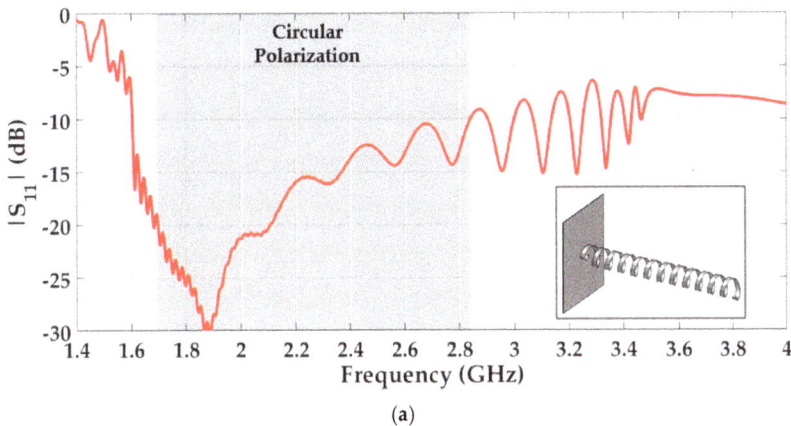

**Figure 7.** *Cont.*

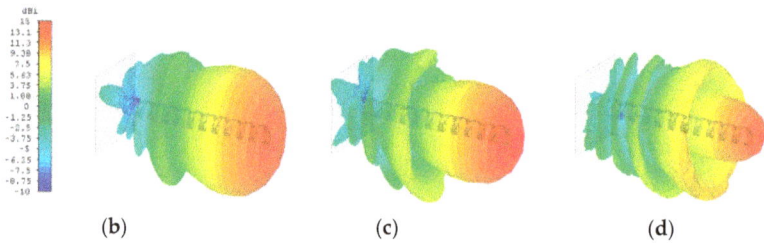

**Figure 7.** Simulated results of a conventional helix antenna in axial mode: (**a**) $|S_{11}|$ results; (**b**) 3D radiation pattern in axial mode at 2 GHz (**c**) 2.45 GHz and (**d**) 2.8 GHz. The conventional helix unit cell dimensions are: $r$ = 19.5 mm, $\alpha$ = 15°, $p$ = 32.82 mm, $w = p/3$ and $L_p$ = 126.84 mm.

The conventional helix antenna has an impedance bandwidth (−10 dB of the $|S_{11}|$) of 54.5%, and an AR bandwidth (below 3dB) of 49%. Figure 7b–d illustrate the 3D radiation pattern of the helix antenna at different frequencies. An end-fire radiation pattern with a high directivity is achieved, as expected for the axial mode.

### 3.2. Twist-And-Glide Symmetrical Helix Antenna Design

The use of the twist-and-glide symmetrical configuration described in Section 2.3.2 can be used to miniaturize a helical antenna due to their ability to increase the propagation constant. The higher the propagation constant, the larger the effect of the miniaturization. Thus, using the twist-and-glide cell configuration illustrated in Figure 5a, choosing a long corrugation length value, the size reduction of the helix antenna can be remarkable.

The twist-and-glide symmetrical helix antenna has the same feed, ground plane and number of turns than the conventional version described in Section 3.1. The only difference between both antennas is the helical strip radius. The twist-and-glide symmetrical helix antenna radius is fixed to 14.8 mm in order to operate with the axial mode. According to Equation (1) and maintaining the pitch angle at 15 degrees, if the helix antenna radius $r$ is smaller, the pitch period $p$ must be smaller, reducing the length of the antenna. Therefore, this new antenna will have a reduced volume, since its length and radius are both smaller. Again, these dimension values for the unit cell imply that the third mode is the one excited in our miniaturized antenna design.

Figure 8a illustrates the simulated reflection coefficient, including the circular polarization bandwidth. A tapered twist-and-glide unit cell is needed in the first turn of the helix antenna for impedance matching.

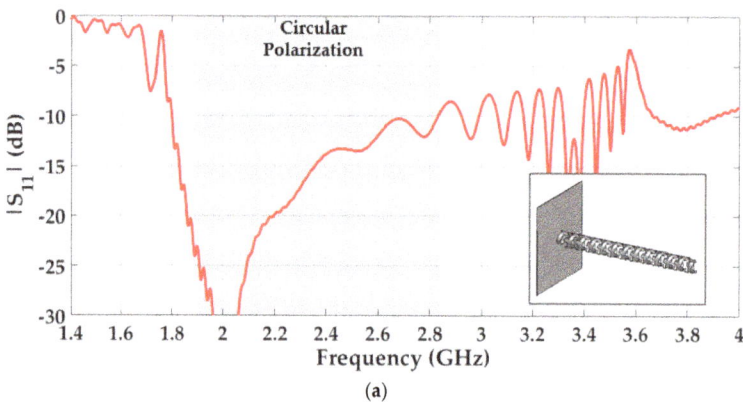

**Figure 8.** *Cont.*

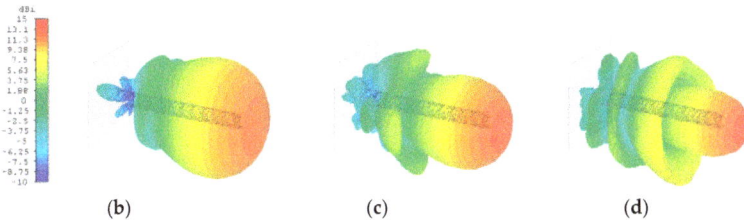

**Figure 8.** Simulated results of the miniaturized glide- and twist-symmetrical helix antenna centered at 2.45 GHz: (**a**) Simulated $|S_{11}|$ (**b**) 3D radiation pattern in axial mode at 2 GHz (**c**) 2.45 GHz and (**d**) 2.8 GHz. The twist-and-glide symmetrical unit cell dimensions are: $r$ = 14.8 mm, $\alpha$ = 15°, $p$ = 24.91 mm, $w = p/3$, $L_p$ = 96.27 mm, $N_{subcell}$ = 4, $h_c$ = 7.5 mm and $w_c = 0.3 L_p / N_{subcell}$.

The simulation results show an impedance bandwidth of 43.4%. Circular polarization is achieved and the AR bandwidth covers 38% maintaining the central frequency at 2.45 GHz. These bandwidths are slightly narrower than in the case of the conventional helix antenna. The loss in bandwidth in the miniaturized version is caused by the difficulty of matching the corrugated strips. The directivity and size comparisons are illustrated in Figure 9a,b. Both directivity levels are above 12 dBi but the directivity of the miniaturized design is lower due to a smaller antenna's physical aperture. Regarding the size comparison, the volume size of the conventional helix antenna has been reduced in a 42.2%.

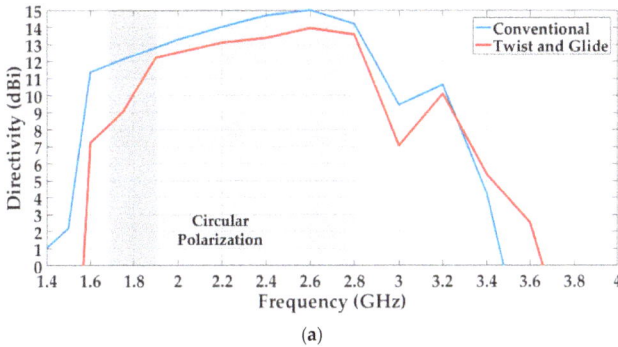

(**a**)

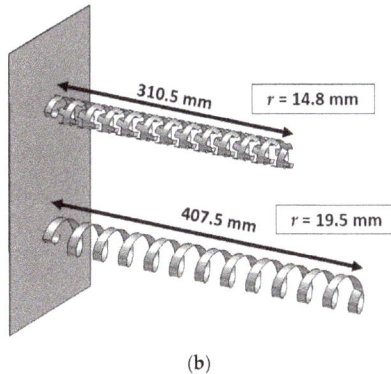

(**b**)

**Figure 9.** (**a**) Directivity comparison (**b**) Antenna size comparison. The twist-and-glide symmetrical unit cell dimensions are: $r$ = 14.8 mm, $\alpha$ = 15°, $p$ = 24.91 mm, $w = p/3$, $L_p$ = 96.27 mm, $N_{subcell}$ = 4, $h_c$ = 7.5 mm and $w_c = 0.3 L_p / N_{subcell}$.

## 4. Conclusions

In this document, we introduce the use of higher symmetries for helix antenna design. Glide and twist symmetries can be employed to modify the propagation properties of periodic structures by increasing the value of the equivalent refractive index. This effect has been employed here to reduce the size of a helix antenna.

With the use of both twist and glide symmetries, the helix achieves a higher propagation constant of the helix, reducing its operational frequency and yielding a miniaturization in the structure. This effect is mainly controlled by the length of the added corrugations. Here, we have demonstrated a reduction of 42.2% in the antenna volume for a helix antenna operating at 2.45 GHz. Our miniaturized helix design achieves similar antenna performance to a conventional helix design in terms of directivity and AR bandwidth.

**Author Contributions:** P.P. and O.Q.-T. conceived the concept. Á.P.-C. designed the helix structures and carried out the simulations. A.A.-A. supported with the preparation of the figures. P.P. and J.V.-V. supervised the work. All authors discussed the content and reviewed and edited the manuscript.

**Funding:** This research was funded by the Spanish Ministerio de Ciencia, Innovación y Universidades with European Union FEDER funds, grant number TIN2016-75097-P, and by Universidad de Granada, under the project PPJI2017.15. Also, this work has been partially supported by the Universidad de Granada through the grant program "Becas de iniciación a la investigación" from Plan Propio de Investigación.

**Conflicts of Interest:** The authors declare no conflict of interest.

## References

1. Rosen, J. *Symmetry in Science: An Introduction to the General Theory*; Springer: Berlin/Heidelberg, Germany, 1995.
2. Courtney, T.H. *Mechanical Behavior of Materials*, 2nd ed.; McGraw-Hill: New York, NY, USA, 2005.
3. Crepeau, P.J.; McIsaac, P.R. Consequences of symmetry in periodic structures. *Proc. IEEE* **1964**, *52*, 33–43. [CrossRef]
4. Hessel, A.; Chen, M.H.; Li, R.C.; Oliner, A.A. Propagation in periodically loaded waveguides with higher symmetries. *Proc. IEEE* **1973**, *61*, 183–195. [CrossRef]
5. Quevedo-Teruel, O.; Ebrahimpouri, M.; Ghasemifard, F. Lens Antennas for 5G Communications Systems. *IEEE Commun. Mag.* **2018**, *56*, 36–41. [CrossRef]
6. Quevedo-Teruel, O.; Ebrahimpouri, M.; Kehn, M.N.M. Ultrawideband metasurface lenses based on off-shifted opposite layers. *IEEE Antennas Wirel. Propag. Lett.* **2016**, *15*, 484–487. [CrossRef]
7. Chen, Q.; Ghasemifard, F.; Valerio, G.; Quevedo-Teruel, O. Modeling and Dispersion Analysis of Coaxial Lines with Higher Symmetries. *IEEE Trans. Microw. Theory Tech.* **2018**, *66*, 4338–4345. [CrossRef]
8. Ebrahimpouri, M.; Quevedo-Teruel, O.; Rajo-Iglesias, E. Design Guidelines for Gap Waveguide Technology Based on Glide-Symmetric Holey Structures. *IEEE Microw. Wirel. Compon. Lett.* **2017**, *27*, 542–544. [CrossRef]
9. Padilla, P.; Herran, L.F.; Tamayo-Dominguez, A.; Valenzuela-Valdes, J.F.; Quevedo-Teruel, O. Glide Symmetry to Prevent the Lowest Stopband of Printed Corrugated Transmission Lines. *IEEE Microw. Wirel. Compon. Lett.* **2018**, *28*, 750–752. [CrossRef]
10. Ghasemifard, F.; Norgren, M.; Quevedo-Teruel, O. Twist and Polar Glide Symmetries: An Additional Degree of Freedom to Control the Propagation Characteristics of Periodic Structures. *Sci. Rep.* **2018**, *8*, 11266. [CrossRef]
11. Quevedo-Teruel, O.; Dahlberg, O.; Valerio, G. Propagation in Waveguides with Transversal Twist-Symmetric Holey Metallic Plates. *IEEE Microw. Wirel. Compon. Lett.* **2018**, *28*, 858–860. [CrossRef]
12. Dahlberg, O.; Mitchell-Thomas, R.C.; Quevedo-Teruel, O. Reducing the Dispersion of Periodic Structures with Twist and Polar Glide Symmetries. *Sci. Rep.* **2017**, *7*, 10136. [CrossRef]
13. Quevedo-Teruel, O.; Miao, J.; Mattsson, M.; Algaba-Brazalez, A.; Johansson, M.; Manholm, L. Glide-Symmetric Fully Metallic Luneburg Lens for 5G Communications at Ka-Band. *IEEE Antennas Wirel. Propag. Lett.* **2018**, *17*, 1588–1592. [CrossRef]
14. Ebrahimpouri, M.; Rajo-Iglesias, E.; Sipus, Z.; Quevedo-Teruel, O. Cost-effective gap waveguide technology based on glide-symmetric holey EBG structures. *IEEE Trans. Microw. Theory Tech.* **2018**, *66*, 927–934. [CrossRef]

15. Rajo-Iglesias, E.; Ebrahimpouri, M.; Quevedo-Teruel, O. Wideband Phase Shifter in Groove Gap Waveguide Technology Implemented with Glide-Symmetric Holey EBG. *IEEE Microw. Wirel. Compon. Lett.* **2018**, *28*, 476–478. [CrossRef]

16. Ebrahimpouri, M.; Brazalez, A.A.; Manholm, L.; Quevedo-Teruel, O. Using glide-symmetric holes to reduce leakage between waveguide flanges. *IEEE Microw. Wirel. Compon. Lett.* **2018**, *28*, 473–475. [CrossRef]

17. Kraus, J.D. The Helical Antenna. *Proc. IRE* **1949**, *37*, 263–272. [CrossRef]

18. Kraus, J.D.; Marhefka, R.J. *Antennas for All Applications*, 3rd ed.; McGraw-Hill: New York, NY, USA, 2002.

19. Tang, X.; Feng, B.; Long, Y. The Analysis of a Wideband Strip-Helical Antenna with 1.1 Turns. *Int. J. Antennas Propag.* **2016**, *2016*, 5950472. [CrossRef]

20. Balanis, C.A. *Antenna Theory: Analysis and Design*, 3rd ed.; Wiley-Interscience: Hoboken, NJ, USA, 2005; pp. 566–573.

21. Rabemanantsoa, J.; Sharaiha, A. Size Reduced Multi-Band Printed Quadrifilar Helical Antenna. *IEEE Trans. Antennas Propag.* **2011**, *59*, 3138–3143. [CrossRef]

22. Takacs, A.; Aubert, H.; Belot, D.; Diez, H. Miniaturisation of quadrifilar helical antenna: Impact on efficiency and phase centre position. *IET Microw. Antennas Propag.* **2013**, *7*, 202–207. [CrossRef]

23. Byun, G.; Choo, H.; Kim, S. Design of a Dual-Band Quadrifilar Helix Antenna Using Stepped-Width Arms. *IEEE Trans. Antennas Propag.* **2015**, *63*, 1858–1862. [CrossRef]

24. Kazemi, R.; Palmer, J.; Quaiyum, F.; Fathy, A.E. Steerable miniaturised printed quadrifilar helical array antenna using digital phase shifters for BGAN/GPS applications. *IET Microw. Antennas Propag.* **2018**, *12*, 1196–1204. [CrossRef]

25. Shi, Y.; Whites, K.W. Miniaturization of helical antennas using dielectric loading. In Proceedings of the IEEE-APS Topical Conference on Antennas and Propagation in Wireless Communications (APWC), Palm Beach, Netherlands Antilles, 3–9 August 2014; pp. 163–166.

26. Neveu, N.; Hong, Y.; Lee, J.; Park, J.; Abo, G.; Lee, W.; Gillespie, D. Miniature Hexaferrite Axial-Mode Helical Antenna for Unmanned Aerial Vehicle Applications. *IEEE Trans. Magn.* **2013**, *49*, 4265–4268. [CrossRef]

27. Chew, D.K.C.; Saunders, S.R. Meander line technique for size reduction of quadrifilar helix antenna. *IEEE Antennas Wirel. Propag. Lett.* **2002**, *1*, 109–111. [CrossRef]

28. Tawk, Y.; Chahoud, M.; Fadous, M.; Costantine, J.; Christodoulou, C.G. The Miniaturization of a Partially 3-D Printed Quadrifilar Helix Antenna. *IEEE Trans. Antennas Propag.* **2017**, *65*, 5043–5051. [CrossRef]

*symmetry*

MDPI

*Article*

# Modeling of Glide-Symmetric Dielectric Structures

Zvonimir Sipus * and Marko Bosiljevac

Faculty of Electrical Engineering and Computing, University of Zagreb, Unska 3, 10000 Zagreb, Croatia;
marko.bosiljevac@fer.hr
* Correspondence: zvonimir.sipus@fer.hr; Tel.: +385-1-6129-798

Received: 6 May 2019; Accepted: 12 June 2019; Published: 18 June 2019

**Abstract:** Recently, there has been an increased interest in exploring periodic structures having higher symmetry properties, primarily based on metallic realization. The design of dielectric glide-symmetric structures has many challenges, and this paper presents a systematic analysis approach based on Floquet mode decomposition and mode matching technique. The presented procedure connects the analysis of standard periodic structures and glide-symmetric realizations, thus giving insight into the wave propagation and interaction characteristics. The obtained results were verified in comparison with results from known references and using a commercial solver, proving that the proposed analysis technique is inherently accurate, and the degree of accuracy depends only on the number of modes used. The proposed analysis approach represents the first step in the design process of dielectric periodic structures with glide symmetry.

**Keywords:** higher symmetries; glide symmetry; periodic structures; dispersion analysis; mode matching

## 1. Introduction

Interest in different artificial electromagnetic structures has led to investigation of various structures, applications and ideas in general. A very promising direction in these developments is structures with higher symmetries. For a periodic structure we can say that it possesses a higher symmetry when the unit cell coincides with itself after more than one geometrical operation of a different kind—translation, rotation, mirroring, etc. [1–3]. The glide operator, for example, is a geometrical transformation composed by a translation and a mirroring with respect to the so-called glide plane. The idea to combine different types of symmetry in a periodic structure will have a strong influence on the propagation properties. One of the most interesting is the possibility to tune the dispersion of lower propagating modes depending on the application, i.e., dispersion can be reduced or increased. In addition, as a direct consequence of this, the electromagnetic bandgap can be extended or reduced, or even completely removed.

Periodic structures with higher symmetry were first studied in the 1960s and 1970s in relation to one-dimensional periodic waveguides [1–3]. Recent work on structures with higher symmetries (in both one and two dimensions) has demonstrated various effects and devices, such as ultra-wideband Luneburg lenses [4,5], leaky-wave antennas with low frequency dependency [6], cost-efficient gap waveguide technology [7,8], contactless flanges with low leakage [9], low-dispersive propagation in periodic structure-based transmission lines [10,11], and fully metallic reconfigurable filters and phase shifters [12]. These results were obtained using metallic structures, however, in many applications only dielectric types of materials are allowed. Therefore, our interest lies in the investigation of the fully dielectric structures with higher symmetry.

Dielectric waveguides containing periodic variation along the propagation direction have been used in many applications ranging from microwave to optical frequencies. They are designed to either support a bounded propagating wave (in microwave or optical filters and distributed feedback reflectors for high-quality lasers) or an unbounded leaky wave (in leaky-wave antennas and optical couplers) [13]. Recently, some new applications were investigated, like plasmonic optical modulators for nanophotonic architectures in which the modulator transmittance is changed with bias voltage [14].

The reported analysis methods were often based on the assumption that periodic variation acts only as a small perturbation of a planar multilayer waveguide [15–17], which may produce erroneous results in many cases, e.g., if the corrugated grating is thick. The proposed rigorous solution is based on the Floquet mode decomposition and a mode matching (MM) technique [18,19] in which the corrugated layer is modeled as a periodic array of infinite dielectric slabs [20]. All previously considered structures contained one layer of corrugations, except in Reference [21] where two layers of corrugations with different periodicities were investigated. The Floquet mode expansion approach can also be combined with the boundary integral formulation in which the boundary conditions at each interface inside the structure follow Floquet's theorem [22]. Similarly, the finite element method can be adapted for computation of modal decomposition and scattering matrices [23,24].

In this paper we have extended the MM analysis method to glide-symmetric dielectric periodic structures. The presented analysis description is focused on one-dimensional glide-symmetric structures [25]. The first part of the paper describes the analysis procedure and it is followed by initial results obtained for a test case found in literature, while the properties of the glide-symmetric dielectric structures are discussed in the second part of the paper.

## 2. Analysis of Waveguides Containing Periodic Dielectric Structures

A periodic structure possesses a higher symmetry when the unit cell coincides with itself after more than one geometrical operation. Glide symmetry is the invariance of a periodic structure under a translation of half its period and a mirroring with respect to a plane parallel to the periodicity directions. Written as a formula, if the periodicity is present along the $x$-direction with the period $P_x$, a glide operator can be written as:

$$\begin{cases} (x,y) \rightarrow (x + \frac{P_x}{2}, y) \\ z \rightarrow -z \end{cases} . \tag{1}$$

Other kinds of higher symmetries are defined by combination of rotation and translation (twist or polar glide symmetries) or by time operations (parity-time symmetries).

The analysis of guiding dielectric structures having higher symmetries is based on Floquet mode decomposition and MM approach [26–29]. The electric and magnetic fields in each section of the structure are represented as a sum of suitable modes with unknown complex amplitude, i.e., we used the preknowledge about the electromagnetic (EM) field configuration and symmetry properties to reduce the number of unknowns, to precisely describe the EM field present in the structure, and to give a physical insight about the presence of higher symmetry properties.

The dielectric waveguide structure of interest is shown in Figure 1. It is a parallel-plate waveguide (PPW) with top and bottom walls realized using dielectric corrugations. The period of corrugations is denoted by $P_x$, the height of corrugations by $h_{corr}$, the width and permittivity of periodic dielectric inclusions by $W_{x1}$, $W_{x2}$ and $\varepsilon_1$, $\varepsilon_2$, and the height and permittivity of parallel-plate waveguide by $h_{ppw}$ and $\varepsilon_{ppw}$. Note that the plane $z = 0$ is located in the middle of the parallel-plate region. We will first analyze the simple periodic structure (i.e., the one with classical translation/mirroring symmetry, see Figure 1a) and then we will modify the analysis procedure for structures with higher symmetry in which the corrugations are shifted with respect to each other by $P_x/2$, i.e., the structure will be glide-symmetric, as in Figure 1b.

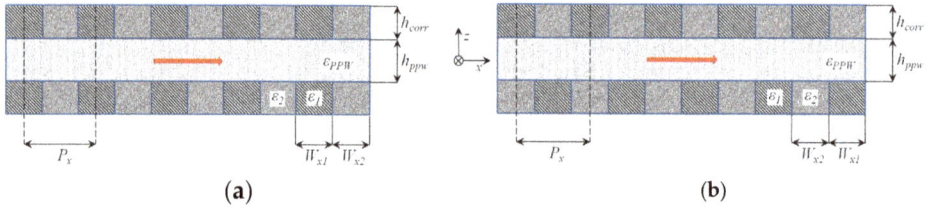

**Figure 1.** Sketch of the parallel-plate dielectric waveguide with periodic structure; (**a**) simple periodic structure with translation and mirroring symmetry, and (**b**) periodic structure with glide symmetry. Direction of wave propagation is sketched using a red arrow.

The waveguide mode of interest is propagating in the direction perpendicular to corrugations (i.e., in the *x*-direction). Therefore, we can distinguish transverse electric (TE) and transverse magnetic (TM) propagating modes (i.e., they do not mix due to presence of the periodic structure corrugations). Here we will give the formulation for TE modes because it is straightforward to repeat the formulation for TM modes.

The analysis is based on representation of the EM field in all regions of interest: Parallel-plate region, layer with corrugations, and outer space. The analysis procedure can be simplified if the modes are orthogonal. This is enforced in the parallel-plate waveguide region by representing the EM field as a series of Floquet harmonics by virtue of periodicity. For each Floquet mode present in the parallel-plate waveguide the *y*-component of the E-field can be expressed as

$$E_y(x,z) = \sum_{m=-N_{PPW}}^{N_{PPW}} \left[ A_m^1 \cos(k_z z) + A_m^2 \sin(k_z z) \right] e^{-jk_x x} \tag{2}$$

$$k_x = k_{x,0} + \frac{2m\pi}{P_x}, \quad k_y = 0, \quad k_z = \sqrt{k_0^2 \varepsilon_{PPW} - k_x^2 - k_y^2}. \tag{3}$$

The tangential *x*-component of *H*-field can be calculated using the following expression

$$H_x = \frac{-j}{\eta_0 k_0} \frac{\partial E_y}{\partial z} = \frac{-j}{\eta_0 k_0} \sum_{m=-N_{PPW}}^{N_{PPW}} k_z \left[ -A_m^1 \sin(k_z z) + A_m^2 \cos(k_z z) \right] e^{-jk_x x}. \tag{4}$$

Here, *m* is the index of the Floquet mode, $N_{PPW}$ is the highest-order considered Floquet mode (in total $2N_{PPW} + 1$ modes are taken into account), and $k_0$ and $\eta_0$ are the wave number and the wave impedance of free space.

In order to determine the Floquet coefficients $A_m$ and the propagation constant $k_{x,0}$ of the propagating wave inside the parallel-plate waveguide we need to match the tangential EM components with the ones in the corrugated walls. The wave propagating in the corrugated region can be modeled as a wave propagating along a periodic array of dielectric slabs [18,20] with the following EM field distribution ($N_{corr}$ denotes the highest-order considered mode).

$$E_y^{corr}(x,z) = \sum_{i=1}^{N_{corr}} E_{0,i}^{corr}(x) \left( C_i^+ e^{-j\gamma_i z} + C_i^- e^{+j\gamma_i z} \right) \tag{5}$$

$$E_{0,i}^{corr}(x) = \begin{cases} B_i^1 \cos k_{x,2}(x + P_x/2) + B_i^2 \sin k_{x,2}(x + P_x/2) & -P_x/2 \le x \le -P_x/2 + W_2/2 \\ B_i^3 \cos k_{x,1} x + B_i^4 \sin k_{x,1} x & -W_1/2 \le x \le W_1/2 \\ B_i^5 \cos k_{x,2}(x - P_x/2) + B_i^6 \sin k_{x,2}(x - P_x/2) & P_x/2 - W_2/2 \le x \le P_x/2 \end{cases} \tag{6}$$

$$H_x^{corr} = \frac{-j}{\eta_0 k_0} \frac{\partial E_y^{corr}}{\partial z}, \quad k_{x,1} = \sqrt{k_0^2 \varepsilon_1 - \gamma_i^2}, \quad k_{x,2} = \sqrt{k_0^2 \varepsilon_2 - \gamma_i^2}. \tag{7}$$

The coefficients $B_i^1 - B_i^6$ and the propagation constant in the $z$-direction $\gamma_i$ are determined by considering the wave propagation along the periodic array with an assumed progressive phase delay per unit cell. In more detail, for each possible propagation constant $k_{x,0}$, i.e., for each considered progressive phase delay we need to solve the secondary (local) mode-matching problem. The problem is described with a linear system of four equations representing the continuity of the $E_y$ and $H_z$ field components at two boundaries $x = \pm W_1/2$. Since for each considered case we have six unknown coefficients as seen in Equations (5)–(7), the Floquet theorem is used to express $B_i^5$ and $B_i^6$ using $B_i^1$, $B_i^2$ and assuming progressive phase delay. The determinantal equation resulting from the linear system gives the value of the propagation constant along the interfaces $\gamma_i$ and then it is possible to determine the field distribution of the considered $i$th mode. The details of the formulation are given in Reference [18].

By matching these two expressions (e.g., for the $E$-field at the boundary $z = h_{\text{ppw}}/2$) and testing it with $(1/P_x) \exp(+jk_x x)$, one equation per each Floquet harmonic is obtained (as a consequence of orthogonality of Floquet harmonics):

$$A_m^1 \cos(k_z h_{ppw}/2) + A_m^2 \sin(k_z h_{ppw}/2) = \sum_{i=1}^{N_{corr}} \widetilde{E}_{0,i}^{corr}(m)\left(C_i^+ e^{-j\gamma_i h_{ppw}/2} + C_i^- e^{+j\gamma_i h_{ppw}/2}\right), \tag{8}$$

$$\widetilde{E}_{0,i}^{corr}(m) = \frac{1}{P_x} \int_{-P_x/2}^{P_x/2} E_{0,i}^{corr}(x) e^{+j(k_{x,0}+(2m\pi/P_x))x} dx. \tag{9}$$

We can further simplify Equation (8) by considering only even or odd modes (with respect to the symmetry plane $z = 0$). By doing so, only one coefficient $A_m^1$ or $A_m^2$ is left and we have consequently expressed each Floquet coefficient $A_m$ with the Fourier transformation of the $E$-field distribution at the corrugation boundary. Thereby, only the coefficients $C_i^\pm$ (among all coefficients $A_m$, $B_i$, $C_i$ and $D_m$ that describe the field distribution) are the unknowns.

The goal is to derive one characteristic equation for the propagation constant $k_{x,0}$ of the propagating wave inside the parallel-plate waveguide. Therefore, we should also match the tangential magnetic field $H_x$ at the boundary between parallel-plate waveguide and corrugated region.

$$\frac{-j}{\eta_0 k_0} \sum_{m=-N_{PPW}}^{N_{PPW}} k_z \left[-A_m^1 \sin(k_z z) + A_m^2 \cos(k_z z)\right] e^{-jk_x x}$$
$$= \frac{-1}{\eta_0 k_0} \sum_{i=1}^{N_{corr}} E_{0,i}^{corr}(x) \gamma_i \left(C_i^+ e^{-j\gamma_i h_{ppw}/2} - C_i^- e^{+j\gamma_i h_{ppw}/2}\right) \tag{10}$$

Equation (10) can be simplified using the connection between the coefficients $A_m$ and $C_i^\pm$ (given by Equations (8) and (9)). By multiplying Equation (10) with $\left(E_{0,l}^{corr}(x)\right)^*$, where $*$ refers to the complex conjugate, and integrating over the period we obtain the following linear system of equations whose determinant is the characteristic equation for the propagation constant $k_{x,0}$

$$\left[Y_{l,i}\right]\left[C_i^\pm\right] = [0]. \tag{11}$$

The coefficients of the linear system of equations related to the unknowns $C_i^+$ and $C_i^-$ are respectively equal (e.g., for the even symmetry case)

$$Y_{l,2i-1} = \sum_{m=-N_{PPW}}^{N_{PPW}} jk_z \left[-\tan(k_z h_{ppw}/2)\right] e^{-j\gamma_i h_{ppw}/2} \widetilde{E}_{0,i}^{corr}(m)\left(\widetilde{E}_{0,l}^{corr}(m)\right)^*$$
$$- \gamma_i e^{-j\gamma_i h_{ppw}/2} \frac{1}{P_x} \int_{-P_x/2}^{P_x/2} E_{0,i}^{corr}(x)\left(E_{0,l}^{corr}(x)\right)^* dx \tag{12}$$

$$Y_{l,2i} = \sum_{m=-N_{PPW}}^{N_{PPW}} jk_z \left[-\tan(k_z h_{ppw}/2)\right] e^{+j\gamma_i h_{ppw}/2} \widetilde{E}_{0,i}^{corr}(m) \left(\widetilde{E}_{0,l}^{corr}(m)\right)^*$$
$$+ \gamma_i e^{+j\gamma_i h_{ppw}/2} \frac{1}{P_x} \int_{-P_x/2}^{P_x/2} E_{0,i}^{corr}(x) \left(E_{0,l}^{corr}(x)\right)^* dx \tag{13}$$

Note that the field distributions of the propagating modes along the periodic array of dielectric slabs $E_{0,i}^{corr}(x)$ (i.e., along the corrugations) are not orthogonal.

Since we have two times more unknowns than equations, we have to repeat the procedure for the half-space above the corrugations. The EM field in the upper half-space (i.e., in the space outside the structure) can be represented as:

$$E_y(x,z) = \sum_{m=-N_{PPW}}^{N_{PPW}} D_m^1 e^{-jk_x x} e^{-jk_z^{air}(z - h_{corr} - h_{PPW}/2)} \tag{14}$$

$$H_x = \frac{-j}{k_0 \eta_0} \frac{\partial E_y}{\partial z}, \quad k_x = k_{x,0} + \frac{2m\pi}{P_x}, \quad k_y = 0, \quad k_z^{air} = \sqrt{k_0^2 - k_x^2 - k_y^2}. \tag{15}$$

Here, we have assumed that the space outside the structure is air (or vacuum) with the relative permittivity equal to one. The corresponding coefficients of the linear system of equations related to the unknown $C_i^{\pm}$ are equal

$$Y_{N_{corr}+l,2i-1} = \sum_{m=-\infty}^{\infty} k_z^{air} e^{-j\gamma_i(h_{corr}+h_{ppw}/2)} \widetilde{E}_{0,i}^{corr}(m) \left(\widetilde{E}_{0,l}^{corr}(m)\right)^*$$
$$- \gamma_i e^{-j\gamma_i(h_{corr}+h_{ppw}/2)} \frac{1}{P_x} \int_{-P_x/2}^{P_x/2} E_{0,i}^{corr}(x) \left(E_{0,l}^{corr}(x)\right)^* dx \tag{16}$$

$$Y_{N_{corr}+l,2i} = \sum_{m=-\infty}^{\infty} k_z^{air} e^{+j\gamma_i(h_{corr}+h_{ppw}/2)} \widetilde{E}_{0,i}^{corr}(m) \left(\widetilde{E}_{0,l}^{corr}(m)\right)^*$$
$$+ \gamma_i e^{+j\gamma_i(h_{corr}+h_{ppw}/2)} \frac{1}{P_x} \int_{-P_x/2}^{P_x/2} E_{0,i}^{corr}(x) \left(E_{0,l}^{corr}(x)\right)^* dx \tag{17}$$

It is enough to consider only the upper half of the structure shown in Figure 1a; the lower part is taken into account using even or odd mirroring symmetry of the considered modes.

For the structures with glide symmetry, i.e., if the upper corrugated plate is shifted by half a period in the x-direction, the E-field distributions at the top and bottom corrugation interfaces can be related as:

$$\mathbf{E}\left(x, y, z = -\frac{h_{PPW}}{2}\right) = e^{-jk_{x,0} P_x/2} \, \mathbf{E}\left(x - \frac{P_x}{2}, y, z = +\frac{h_{PPW}}{2}\right). \tag{18}$$

This expression should have the double sign ± in front of the exponential term. However, the double sign ± does not define two different sets of modes or introduce any ambiguity because the solutions of the two problems actually coincide: the Floquet harmonic 0 becomes the harmonic −1 when switching the sign from plus to minus (see Reference [28] for details).

We will use the following translation property of the Fourier transformation:

$$\frac{1}{P_x} \int \varphi(x - P_x/2) \, e^{-jk_{x,0} P_x/2} e^{jk_x x} dx = \frac{1}{P_x} \int \varphi(x') \, e^{+jm\pi} e^{jk_x x'} dx' = \widetilde{\varphi}(k_x) e^{+jm\pi} = \widetilde{\varphi}(k_x) \cdot (-1)^m. \tag{19}$$

Term $(-1)^m$ actually means that depending on the index of Floquet mode we have even or odd symmetry across the z = 0 plane (see Equation (2)). For even Floquet modes ($m = 0, \pm2, \pm4, \ldots$) the E-field in the parallel-plate region is described with $A_m^1 \cos(k_z z)$ terms, while for odd Floquet modes ($m = \pm1, \pm3, \ldots$) the E-field is described with $A_m^2 \sin(k_z z)$ terms. In other words, the presence of glide symmetry causes the odd and even symmetries to be mixed inside the PPW resulting in extraordinary properties of the glide-symmetric guiding structure.

## 3. Results

In order to test the developed code, we compared the results of a calculated field distribution inside the corrugated layer with the results from [18]. We plotted the normalized value of the $H_y$ field component of the TM wave for different values of assumed progressive phase delay per unit cell. The parameters of the corrugated structure are the following: $\varepsilon_{r1} = 2.56$, $\varepsilon_{r2} = 1.0$, $P_x = 0.6 \lambda_0$, $W_{x1} = 0.26 \lambda_0$ and $W_{x2} = 0.34 \lambda_0$. Excellent matching of the obtained results can be noticed in Figure 2. We also tested the accuracy of the calculated propagation constant of the first two modes travelling along the periodic array of dielectric slabs (Figure 3). Again we got excellent agreement with results given in [18]. It is interesting to notice that for values $k_{x,0}$ less than $0.5 \cdot \pi/P_x$ the first high-order mode is evanescent, i.e., the values of $\gamma_2$ are imaginary ($k_{x,0}$ is the transverse propagation constant in the corrugated region). For values $k_{x,0}$ larger than $0.5 \cdot \pi/P_x$ the wave propagates along the dielectric slabs and consequently $\gamma_2$ is a real number. The dominant mode is of a propagating type for all values of $k_{x,0}$.

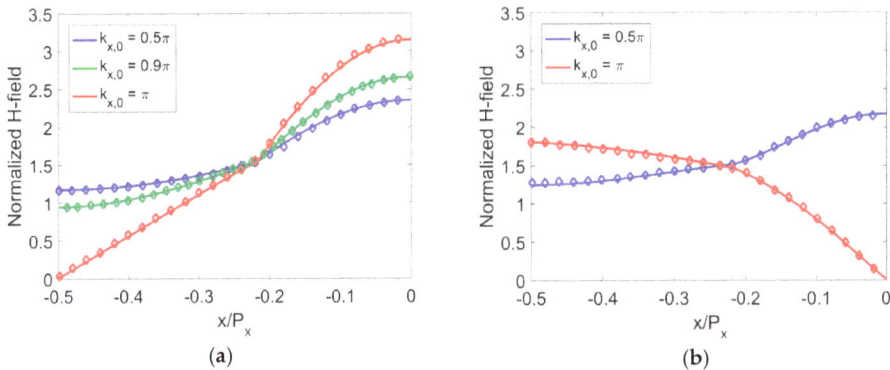

(a)                                    (b)

**Figure 2.** Normalized value of the $H$ field component of TM wave for different values of assumed progressive phase delay per unit cell; solid line—results calculated using the developed program; diamonds—calculated results from [18]; (a) first mode, (b) second mode.

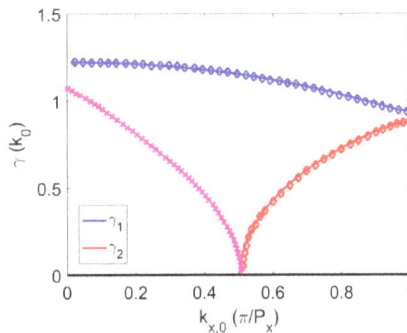

**Figure 3.** Normalized value of the propagation constant of the first two modes travelling along the periodic array of dielectric slabs; solid line—results calculated using the developed program, diamonds—calculated results from Reference [18]. $\gamma_2 = 0$ is the cut-off condition of the 2nd mode, thus for the values $k_{x,0}$ smaller than the cut-off value ($0.5 \cdot \pi/P_x$) $\gamma_2$ is imaginary (the mode is evanescent).

In order to investigate the properties of glide-symmetric dielectric structures we considered the following periodic structure: $\varepsilon_{PPW} = 2.56$, $h_{PPW} = 10$ mm, $P_x = 35$ mm, $W_{x1} = 15$ mm, $W_{x2} = 20$ mm, $\varepsilon_{r1} = 10.0$, $\varepsilon_{r2} = 1.0$, and $h_{corr} = 10$ mm. First, we analyzed the simple periodic structure (Figure 1a). In Figure 4 the dispersion diagram of the first three propagating modes are given. Note that only the values of the propagation constant larger than the free-space propagation constant $k_0$ are given, otherwise the excited mode is a fast wave and the structure is radiating part of the EM energy (i.e., we have a leaky-wave antenna). The obtained results were compared with the ones obtained using a general electromagnetic solver, CST Microwave Studio in our case, and the agreement is very good. Specifically, the Eigenmode Solver in the CST Microwave Studio package was used to find the dispersion characteristics of the considered structures, thus it was necessary to analyze only a unit cell. In $x$- and $y$- directions the periodic boundary conditions were applied, while at the top and bottom of the periodic cell the PMC boundary conditions were used to simulate infinitely long symmetric and glide-symmetric structures. Note that the unit cell was quite long in the $z$-direction since it was needed to ensure that the amplitude of the evanescent fields was negligible at the top and bottom boundaries.

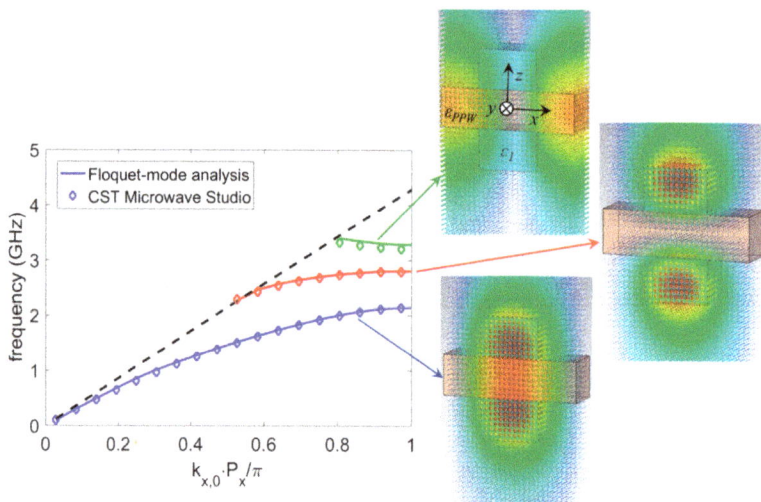

**Figure 4.** Propagation constant of the first three modes travelling along the dielectric waveguide with corrugations from Figure 1a; solid line—results calculated using the developed program (blue line: 1st even mode, red line: 1st odd mode, green line: 2nd even mode), diamonds—results calculated using CST Microwave Studio (dashed black line represents the light-line). The E-field distribution in the unit cell of the first three modes for $k_{x,0} = \pi/P_x$ is also shown (the structure is infinite in the $y$-direction).

In Figure 4 the field distributions of the first three propagating modes are also given. Note that the dominant mode is even (relative to the symmetry line $z = 0$), the first higher order-mode is odd, and the second higher-order mode is even again. Note also that the evanescent field is present also in the air close to the structure. This is probably the most important difference compared to the metallic waveguides. The higher the mode (or for frequencies closer to the cut-off frequency), the larger the percentage of the power actually propagating out of the parallel-plate waveguide and consequently in the half-spaces around the guiding structure. In Figure 4, the third mode has more than half of the power propagating outside the waveguiding structure. Theoretically, for frequencies just above cut-off almost all power is actually propagating outside the parallel-plate layer [30]. This is one of the limits in the design of dielectric glide-symmetric structures since the designed metallic components mostly have a very thin parallel-plate region (in order to obtain the desired electromagnetic properties).

In the dielectric case probably there is no point to make equivalent structures with a thin gap between corrugations since almost all power will propagate outside the guiding structure.

The dispersion diagram of the glide-symmetric dielectric waveguide (having the same dimensions as considered simple periodic structure) is given in Figure 5. The shape of the obtained dispersion diagram is similar to the ones obtained with metallic glide-symmetric waveguides. The obtained bandgap is between 2.52 and 3.65 GHz. The field distribution of the first two modes is also shown for $k_{x,0} = \pi/P_x$. Like in the case of the simple periodic structure (Figure 4) a much larger percentage of EM power is propagating outside the dielectric structure for the second mode. The selected time moment gives the best illustration of the E-field distribution for the depicted modes. Other time moments could be selected, as illustrated in Figure 6 where the E-field distribution of the dominant mode is given for relative phase shifts 45°, 90°, 135° and 180°, respectively.

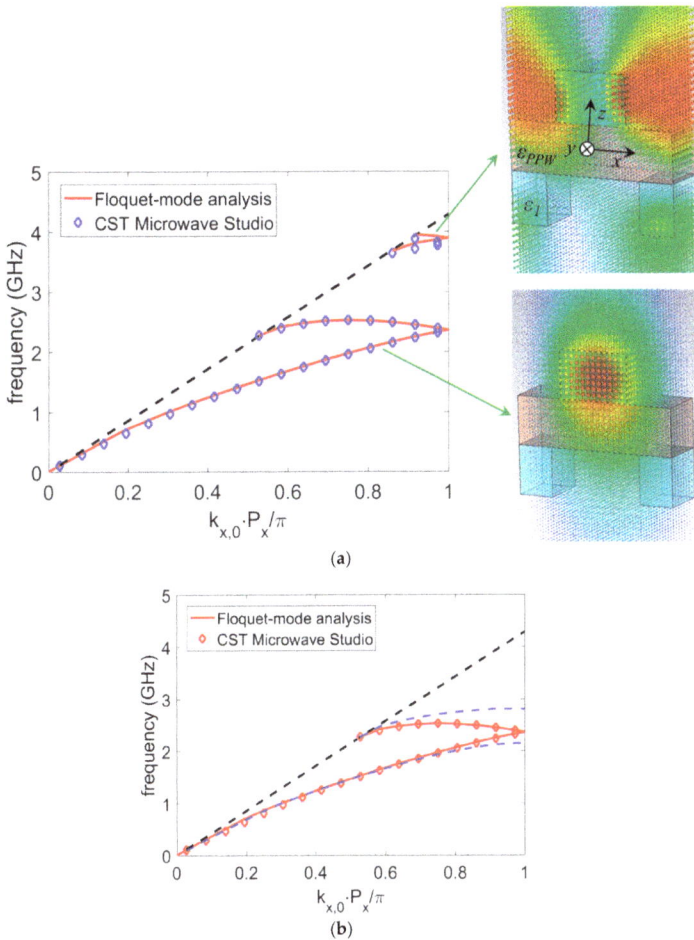

**Figure 5.** Propagation constant of the first two modes travelling along the glide-symmetric dielectric waveguide with corrugations (Figure 1b); solid line—results calculated using the developed program, diamonds—results calculated using CST Microwave Studio (dashed black line represents the light-line); (**a**) glide-symmetric structure only, (**b**) comparison of glide-symmetric and simple-symmetric structures (blue dashed line). The E-field distribution in the unit cell of the first two modes for $k_{x,0} = \pi/P_x$ is also shown (the structure is infinite in the $y$-direction).

| $t = T/8$ | $t = T/4$ | $t = 3T/8$ | $t = T/2$ |

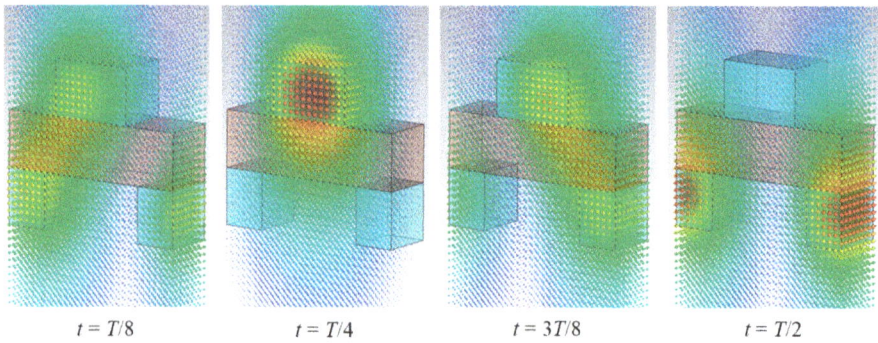

**Figure 6.** The E-field distribution in the unit cell of the first mode for $k_{x,0} = \pi/P_x$ at different time moments where $T$ denotes the period of oscillations ($T = 1/f$).

In order to be able to tune the dispersion diagram according to the designer's needs one should understand the background of such a shape. The key point is Equation (19), i.e., the fact that glide symmetry actually means mixing of odd and even modes. This is visible in Figure 5b where we have displayed on the same diagram dispersion curves of simple and glide-symmetric dielectric waveguides. It can be seen that the lower part of the dispersion curve is nearly indistinguishable from the even mode profile of the non-glide structure (up to the vicinity of the propagation constant $k_{x,0} = \pi/P_x$), while the upper part is mostly affected by the odd mode.

This property actually enabled us to tune the dispersion diagram (see Figure 7). For example, if we reduced the thickness of the parallel-plate region from 10 mm to 2 mm, then the odd mode had a higher cut-off frequency and therefore the dispersion diagram of a glide-symmetric case was less dispersive and the bandgap started at higher frequencies. This was even more prominent if we reduced the thickness of the corrugated region from 10 mm to 2 mm (Figure 7b) because the overall thickness of the dielectric structure was then much smaller.

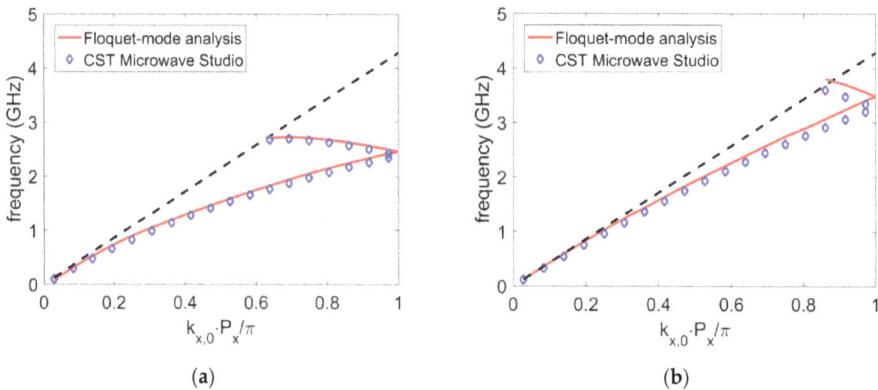

(a)

(b)

**Figure 7.** Propagation constant of the first mode travelling along the glide-symmetric dielectric waveguide with corrugations (case of thinner parallel-plate dielectric slab); solid line—results calculated using the developed program, diamonds—results calculated using CST Microwave Studio (dashed black line represents the light-line); (**a**) case of thinner parallel-plate dielectric slab ($h_{ppw} = 2$ mm), (**b**) case of thinner corrugated region ($h_{corr} = 2$ mm).

We tested various combinations of refractive indexes for PPW and corrugated regions. One natural choice is to have only one material, i.e., that corrugations are made from the same material as the

PPW ($\varepsilon_{r1} = \varepsilon_{PPW} = 2.56$ and $\varepsilon_{r2} = 1.0$; other parameters are the same as for the structure in Figure 5). In other words, such structures are made from a homogeneous media but the boundaries have a periodic variation. Results of the same type were obtained (see Figure 8), however the glide-symmetric properties were not so pronounced. It can be concluded that contrast between the permittivity of PPW and corrugated regions can also be used for tuning the dispersion properties.

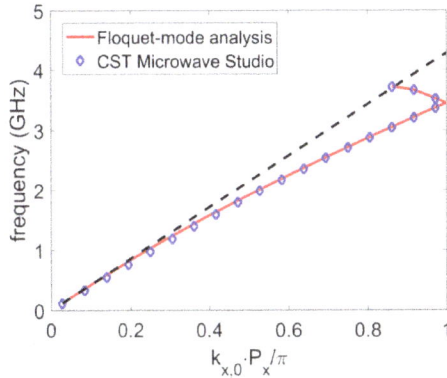

**Figure 8.** Propagation constant of the first mode travelling along the glide-symmetric dielectric waveguide with corrugations made from homogeneous material; solid line—results calculated using the developed program, diamonds—results calculated using CST Microwave Studio (dashed black line represents the light-line).

Needless to say, it is possible to analyze guiding structures in which the lower corrugated layer is arbitrarily shifted compared to the upper layer (i.e., when the shift is between 0 and $P_x/2$). The obtained dispersion diagrams are located between ones for the symmetric and glide-symmetric cases, which was demonstrated in Reference [11] for metallic structures. Therefore, we have focused our investigation on the two most interesting cases, symmetric and glide-symmetric.

Finally, numerical properties of the proposed analysis method should be mentioned. The number of modes in the PPW and corrugated regions were $N_{PPW} = 5$ (in total 11 modes) and $N_{corr} = 2$. We tested the numerical convergence of the solution, and enlarging the number of modes only slightly improved the accuracy of the solution. The needed computer time was less than a second per point in the dispersion diagram, while the CST typically needed around 1 hour for calculating the whole dispersion diagram (depending on the accuracy of the meshing). All the calculations were made on a standard PC.

## 4. Conclusions

This paper discussed the Floquet mode decomposition applied together with a mode-matching approach in the analysis and design of dielectric waveguides with glide-symmetric periodicity. The analysis method is based on representing the EM field in each region with suitable modes and connecting the field distributions in different regions using symmetry properties, resulting in an efficient program for determining propagation properties and obtaining a physical picture of modes in such waveguides. The developed approach provides a basis for better understanding of wave propagation characteristics and of various types of wave interactions in glide-symmetric structures.

Until now most realized waveguide prototypes possessing higher symmetry were made from metal. Although at first glance there are a lot of similarities between metallic and dielectric glide-symmetric structures, the main difference comes from the fact that in the dielectric case, part of the propagating wave (and thus part of the electromagnetic power) travels outside the dielectric parallel-plate waveguide. Therefore, the dispersion diagram (and consequently the bandgap) is dominantly conditioned by the odd mode of the equivalent non-glide structure whose cut-off frequency mostly depends on the total

*Symmetry* **2019**, *11*, 805

thickness and permittivity of the waveguide structure. This also limits us in selecting the thickness of the parallel-plate waveguide since the percentage of power propagating outside the guiding structure is larger for thinner structures. The other limit represents the free-space wavenumber—for propagation constants smaller than the free-space wavenumber a fast wave is excited which leads to leakage of electromagnetic energy. Our presented discussion shows that by adjusting the structure parameters various waveguiding properties can be obtained, and our formulation is able to aid in the design process and provide fast and accurate results.

**Author Contributions:** The contributions of both authors were approximately the same. Both authors worked together to develop the present manuscript.

**Funding:** This work was funded by Croatian Science Foundation under the project IP-2018-01-9753.

**Conflicts of Interest:** The authors declare no conflict of interest.

## References

1.  Crepeau, P.J.; McIsaac, P.R. Consequences of symmetry in periodic structures. *Proc. IEEE* **1964**, *52*, 33–43. [CrossRef]
2.  Mittra, R.; Laxpati, S. Propagation in a waveguide with glide refection symmetry. *Can. J. Phys.* **1965**, *43*, 353–372. [CrossRef]
3.  Hessel, A.; Chen, M.H.R.; Li, C.M.; Oliner, A.A. Propagation in periodically loaded waveguides with higher symmetries. *Proc. IEEE* **1973**, *61*, 183–195. [CrossRef]
4.  Quevedo-Teruel, O.; Ebrahimpouri, M.; Ng Mou Kehn, M. Ultrawideband Metasurface Lenses Based on Off-Shifted Opposite Layers. *IEEE Antennas Wirel. Propag. Lett.* **2016**, *15*, 484–487. [CrossRef]
5.  Quevedo-Teruel, O.; Miao, J.; Mattsson, M.; Algaba-Brazalez, A.; Johansson, M.; Manholm, L. Glide-Symmetric Fully Metallic Luneburg Lens for 5G Communications at Ka-Band. *IEEE Antennas Wirel. Propag. Lett.* **2018**, *17*, 1588–1592. [CrossRef]
6.  Dahlberg, O.; Pucci, E.; Wang, L.; Quevedo-Teruel, O. Low-Dispersive Glide-Symmetric Leaky-Wave Antenna at 60 GHz. In Proceedings of the 13th European Conference on Antennas and Propagation, Krakow, Poland, 31 March–5 April 2019.
7.  Ebrahimpouri, M.; Quevedo-Teruel, O.; Rajo-Iglesias, E. Design guidelines for gap waveguide technology based on glide-symmetric holey structures. *IEEE Microw. Wirel. Compon. Lett.* **2017**, *27*, 542–544. [CrossRef]
8.  Ebrahimpouri, M.; Rajo-Iglesias, E.; Sipus, Z.; Quevedo-Teruel, O. Cost-effective gap waveguide technology based on glide-symmetric holey EBG structures. *IEEE Trans. Microw. Theory Tech.* **2018**, *66*, 927–934. [CrossRef]
9.  Ebrahimpouri, M.; Algaba-Brazalez, A.; Manholm, L.; Quevedo-Teruel, O. Using glide-symmetric holes to reduce leakage between waveguide flanges. *IEEE Microw. Wirel. Compon. Lett.* **2018**, *28*, 473–475. [CrossRef]
10. Padilla, P.; Herran, L.F.; Tamayo-Dominguez, A.; Valenzuela-Valdes, J.F.; Quevedo-Teruel, O. Glide symmetry to prevent the lowest stopband of printed corrugated transmission lines. *IEEE Microw. Wirel. Compon. Lett.* **2018**, *28*, 750–752. [CrossRef]
11. Quesada, R.; Martín-Cano, D.; García-Vidal, F.J.; Bravo-Abad, J. Deep subwavelength negative-index waveguiding enabled by coupled conformal surface plasmons. *Opt. Lett.* **2014**, *39*, 2990–2993. [CrossRef] [PubMed]
12. Rajo-Iglesias, E.; Ebrahimpouri, M.; Quevedo-Teruel, O. Wideband phase shifter in groove gap waveguide technology implemented with glide-symmetric holey EBG. *IEEE Microw. Wirel. Compon. Lett.* **2018**, *28*, 476–478. [CrossRef]
13. Elachi, C. Waves in active and passive periodic structures: A review. *Proc. IEEE* **1976**, *64*, 1666–1698. [CrossRef]
14. Babicheva, V.E.; Lavrinenko, A.V. Plasmonic modulator optimized by pattering of active layer and tuning permittivity. *Opt. Commun.* **2012**, *285*, 5500–5507. [CrossRef]
15. Hope, L.L. Theory of optical grating couplers. *Opt. Commun.* **1972**, *11*, 2234–2241. [CrossRef]
16. Harris, J.A.; Winn, R.K.; Dalgoutte, D.G. Theory and design of periodic couplers. *Appt. Opt.* **1972**, *11*, 2234–2241. [CrossRef] [PubMed]

17. Stolland, H.; Yariv, A. Coupled-mode analysis of periodic dielectric waveguides. *Opt. Commun.* **1973**, *8*, 5–8. [CrossRef]

18. Peng, S.T.; Tamir, T.; Bertoni, H.L. Theory of periodic dielectric waveguides. *IEEE Trans. Microw. Theory Tech.* **1975**, *23*, 123–133. [CrossRef]

19. Peng, S.T. Rigorous formulation of scattering and guidance by dielectric grating waveguides: General case of oblique incidence. *J. Opt. Soc. Am. A* **1989**, *6*, 1869–1883. [CrossRef]

20. Lewis, L.R.; Hessel, A. Propagation characteristics of periodic arrays of dielectric slabs. *IEEE Trans. Microw. Theory Tech.* **1971**, *19*, 276–286. [CrossRef]

21. Peng, S.T. Rigorous analysis of guided waves in doubly periodic structures. *J. Opt. Soc. Am. A* **1990**, *7*, 1448–1456. [CrossRef]

22. Hadjicostas, G.; Butler, J.K.; Evans, G.A.; Carlson, N.W.; Amantea, R. A numerical investigation of wave interaction in dielectric waveguides with periodic surface corrugations. *IEEE J. Quantum Electron.* **1990**, *26*, 893–902. [CrossRef]

23. Bao, G. Finite element approximation of time harmonic waves in periodic structures. *SIAM J. Numer. Anal.* **1995**, *32*, 1155–1169. [CrossRef]

24. Dossou, K.; Byrne, M.A.; Botten, L.C. Finite Element Computation of Grating Scattering Matrices and Application to Photonic Crystal Band Calculations. *J. Comput. Phys.* **2006**, *219*, 120–143. [CrossRef]

25. Valerio, G.; Sipus, Z.; Grbic, A.; Quevedo-Teruel, O. Accurate equivalent-circuit descriptions of thin glide-symmetric corrugated metasurfaces. *IEEE Trans. Antennas Propag.* **2017**, *65*, 2695–2700. [CrossRef]

26. Wexler, A. Solution of waveguide discontinuities by modal analysis. *IEEE Trans. Microw. Theory Tech.* **1967**, *15*, 508–517. [CrossRef]

27. Clarricoats, P.J.B.; Slinn, K.R. Numerical method for the solution of waveguide-discontinuity problems. *Electron. Lett.* **1966**, *2*, 226–228. [CrossRef]

28. Valerio, G.; Ghasemifard, F.; Sipus, Z.; Quevedo-Teruel, O. Glide-Symmetric All-Metal Holey Metasurfaces for Low-Dispersive Artificial Materials: Modeling and Properties. *IEEE Trans. Microw. Theory Tech.* **2018**, *66*, 3210–3223. [CrossRef]

29. Ghasemifard, F.; Norgren, M.; Quevedo-Teruel, O.; Valerio, G. Analyzing Glide-Symmetric Holey Metasurfaces Using a Generalized Floquet Theorem. *IEEE Access* **2018**, *6*, 71743–71750. [CrossRef]

30. Okamoto, K. *Fundamentals of Optical Waveguides*, 2nd ed.; Academic Press: Cambridge, MA, USA, 2006.

symmetry

MDPI

*Article*

# Bloch Analysis of Electromagnetic Waves in Twist-Symmetric Lines

Mohammad Bagheriasl * and Guido Valerio *

Laboratoire d'Électronique et Électromagnétisme, Sorbonne Université, F-75005 Paris, France
* Correspondence: mohammad.bagheriasl@sorbonne-universite.fr (M.B.);
  guido.valerio@sorbonne-universite.fr (G.V.)

Received: 15 March 2019; Accepted: 23 April 2019; Published: 3 May 2019

**Abstract:** We discuss here under which conditions a periodic line with a twist-symmetric shape can be replaced by an equivalent non-twist symmetric structure having the same dispersive behavior. To this aim, we explain the effect of twist symmetry in terms of coupling among adjacent cells through higher-order waveguide modes. We use several waveguide modes to accurately derive the dispersion diagram of a line through a multimodal transmission matrix. With this method, we can calculate both the phase and attenuation constants of Bloch modes, both in shielded and open structures. In addition, we use the higher symmetry of these structures to further reduce the computational cost by restricting the analysis to a subunit cell of the structure instead of the entire unit cell. We confirm the validity of our analysis by comparing our results with those of a commercial software.

**Keywords:** dispersion analysis; higher symmetry; periodic structures; twist symmetry; transmission matrix

## 1. Introduction

A higher symmetric periodic line is characterized by its invariance under a geometrical transformation other than a translation. The higher symmetry resulting from a combination of a rotation and a translation is called twist (or screw) symmetry [1–4]. In an $N$-fold twist symmetry, if the period is $p$, a subunit cell ($1/N$ of the entire unit cell) is translated $p/N$ along the periodicity axis and twisted $2\pi/N$ radians around the same axis. Figure 1a,b shows 2-fold and 4-fold twist symmetric lines, respectively.

Recent studies on higher-symmetric structures have revealed interesting characteristics such as a reduced dispersion and wider and stronger stop bands [5–12]. These characteristics have led to numerous applications in different areas such as wideband lens antennas [13], cost-effective gap-waveguide technology [14,15], leaky-wave antennas [16], and leakage reduction in waveguide transitions [17]. All these studies use commercial software to perform periodic analyses on a unit cell of periodicity. Despite its accuracy, this approach is not valid if the structure is open (radiating line) and does not give any physical insight about the impact of the twist operation on the periodic line. Only in Reference [5] was an equivalent circuit proposed for a line based on a coaxial cable. However, that circuit considers the interaction between adjacent cells only through the fundamental TEM mode of the coaxial line. In this paper, we will show that such an approach can fail if dense lines are studied.

As proved in those papers, the advantages of higher-symmetric structures over simpler periodic ones stimulate now the urge to understand where the difference between the two categories arises, for instance, to explain the source of different dispersive behaviors of the structures in Figure 1a,c on the basis of fundamental electromagnetic theory. In addition, there is a need for fast and efficient analysis methods for higher symmetric structures. This paper answers both of these needs by first showing the different coupling mechanism occurring in higher-symmetric and non-higher-symmetric structures and then by formulating a Bloch analysis method that only takes into account a partial unit cell, thus accelerating the calculation procedure.

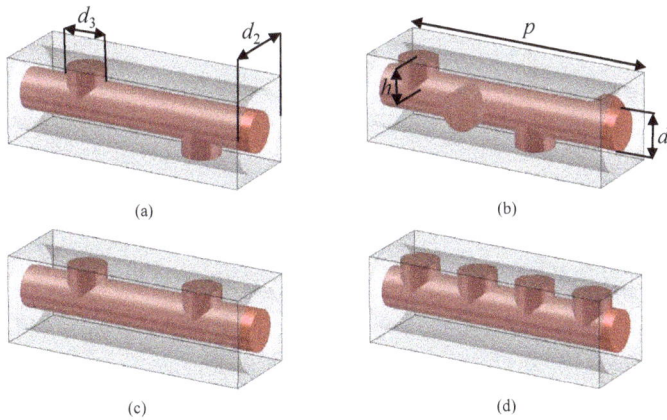

(a)

(b)

(c)

(d)

**Figure 1.** Pin-loaded coaxial transmission lines: (**a**) 2-fold twist-symmetric, (**b**) 4-fold twist-symmetric, (**c**) non-twist symmetric with 2 pins, and (**d**) non-twist symmetric with 4 pins. ($d_1$ = 2.4 mm, $d_2$ = 4.84 mm, $d_3$ = 2.4 mm, and $h$ = 2 mm).

In this work, we first classify periodic structures with twist symmetry into two groups: reducible and irreducible. The former group consists of structures that are equivalent to periodic structures with a reduced period and no twist symmetry, while the latter group consists of those of which the higher symmetry cannot be reduced to an equivalent periodic structure with a reduced period. We show that the couplings between cells through higher-order modes are the cause of the difference between these two groups. We quantify this effect by defining a multimodal transmission matrix to formulate a Bloch analysis. Finally, we formulate, for the first time, the eigenvalue problem based on a twist symmetry operator [4] by means of a transmission-matrix approach. This restricts the problem to a subregion of a unit cell and, therefore, reduces the computational cost of the analysis. It also helps to explain how the higher-order modes affect the reducibility of the structure.

## 2. Reducibility of Twist-Symmetric Structures

In this section, we define the reducibility of twist-symmetric structures to non-twist periodic lines. We will state under which assumptions this property holds by means of a multimodal transmission-matrix approach capable to perform accurate Bloch analyses also for tightly coupled adjacent cells.

### 2.1. Reducible and Irreducible Twist Symmetries

A periodic structure with an $N$-fold twist symmetry is invariant under the twist operator $S_{N,p\hat{z}}$:

$$S_{N,p\hat{z}} : (\rho, \phi, z) \rightarrow (\rho, \phi + \frac{2\pi}{N}, z + \frac{p}{N}) \tag{1}$$

It is easy to observe from Equation (1) that a composition of $N$ twist operators gives $T_{p\hat{z}}$, the translation operator of length $p$ along $z$ ($S_{N,p\hat{z}}^N = T_{p\hat{z}}$). This confirms that a twist-symmetric structure ("twisted structure" in the following) is also a periodic structure. Figure 1a,b shows the unit cells of twisted coaxial transmission lines loaded with circular pins, with a 2-fold and a 4-fold symmetry, respectively.

For each twisted line, we define an associated non-twisted periodic line by suppressing the rotation in Equation (1). The twisted structures in Figure 1a,b are then associated to the non-twisted periodic lines in Figure 1c,d, respectively. While $p$ is the period of the $N$-fold twisted structure, the associated non-twisted structure has a period of $p/N$.

We define "reducible" a twisted structure which has the same dispersion diagram as its corresponding non-twisted one in a certain range of frequencies. Irreducible structures are those without such characteristic. In other words, in a reducible line, the rotations of the subunit cells do not affect the dispersive behavior, while in an irreducible line, the rotations change the dispersion behavior with respect to the associated non-twisted periodic line. Figure 2a shows the comparison between the dispersion diagram of a 2-fold twisted structure as in Figure 1a and its non-twisted associated structure as in Figure 1c. The results are obtained with the CST eigensolver tool (CST ES) [18]. The structure in Figure 1a is reducible to a non-twisted periodic line, since the two dispersion curves are perfectly superimposed. The rotation of its subunit cells does not have an impact on its dispersion diagram. Figure 2b depicts the dispersion diagrams of similar structures as in Figure 1a,c but with different parameters (a shorter period of $p = 7$ mm). This time, the twisted line is irreducible, since it has a different dispersion diagram with respect to the non-twisted one. Figure 2c depicts the dispersion diagrams for a 4-fold twisted structure as in Figure 1b and its associated non-twisted structure as in Figure 1d with $p = 15$ mm. The comparison of the diagrams suggests that the 4-fold twisted line is irreducible at higher frequencies, while it can be reduced to a simple non-twisted structure with a smaller period in the lower frequencies. Figure 2d depicts the dispersion diagrams for the the same structures where the twisted line has a smaller period of $p = 10$ mm. The difference between the two results in this case demonstrates that this 4-fold twisted line is irreducible even at low frequencies.

We want to show that the difference between a reducible and an irreducible twisted line is due to the relevance of higher-order modes of the unloaded line in the coupling between adjacent cells. In order to prove this, we present a rigorous multimodal Bloch analysis and we apply it to different twisted lines.

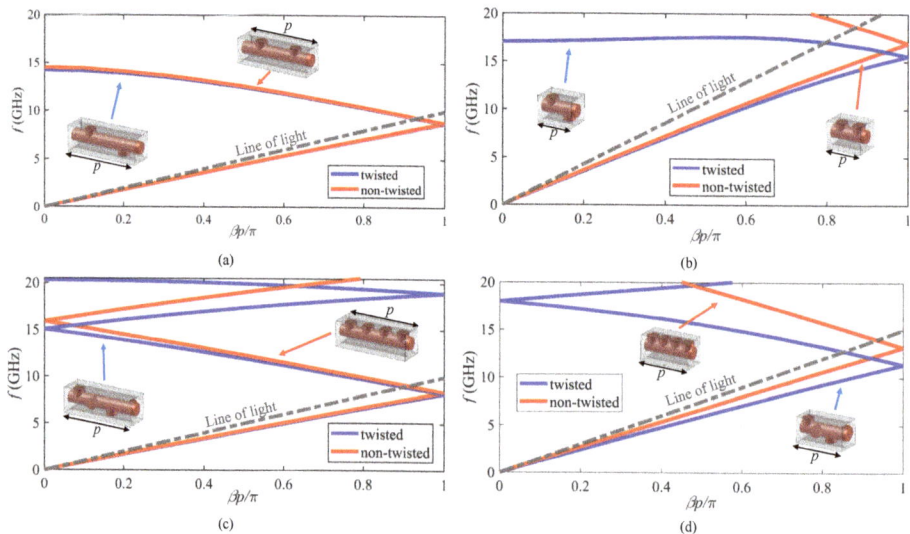

**Figure 2.** A dispersion diagram comparison of the twist-symmetric and their non-twist symmetric structures shown in Figure 1: (**a**) with 2 pins ($p = 15$ mm), (**b**) with 2 pins ($p = 7$ mm), (**c**) with 4 pins ($p = 15$ mm), and (**d**) with 4 pins ($p = 10$ mm).

## 2.2. Multimodal Transmission-Matrix Method

It is well-known that a periodic structure can be studied by limiting the analysis to its unit cell. Considering a 1-D periodic structure with period $p$ along the $z$ direction, Floquet boundary conditions can be written for the unit cell periodic boundaries. This results in an eigenvalue problem for the translation operator $T_{p\hat{z}}$:

$$T_{p\hat{z}}\left[E\left(\rho,\phi,z\right)\right] = E\left(\rho,\phi,z+p\right) = e^{-jk_z p}E\left(\rho,\phi,z\right) \qquad (2)$$

of which the eigenvalue provides the propagation constant of a Bloch mode ($k_z$) and of which the eigenvector is the modal electric field. In general, the propagation constant ($k_z = \beta - j\alpha$) is a complex quantity. Its real part $\beta$ is the phase constant, and the opposite of its imaginary part $\alpha$ is the attenuation constant. An equivalent boundary condition could also be written in terms of the magnetic field $H$. However, we will write all the subsequent conditions in this paper in terms of $E$ for brevity. Monochromatic quantities are assumed throughout the paper, of which the time dependence $e^{j\omega t}$ is suppressed for simplicity.

In order to study the impact of higher-order modes on the cell coupling, we consider on the two periodic boundary planes of a unit cell $n$ modes of the unloaded line ("background modes" in the following). Since the background structure has a circular section, the radial component of the electric field of the background modes has the following profile:

$$\Psi_{(\pm)}^{(m,i),e/h}\left(\rho,\phi\right) = f_{m,i}\left(\rho\right)e^{\pm jm\phi} \qquad (3)$$

where $f_{m,i}$ are suitable linear combinations of Bessel functions expressing the radial dependence of each mode. The index $m$ represents the order of angular dependence ($e^{\pm jm\phi}$), the index $i$ is the radial dependence, and $e$ and $h$ in the superscript stand for TM and TE modes, respectively.

We associate to each mode a voltage and a current. Different definitions are possible for the equivalent transmission lines associated to a waveguide mode, either based on line integrals of the transverse electric and magnetic fields or as quantities proportional to the transverse fields [19]. The method presented here can be used with either one of the formulations as long as a coherent definition is kept throughout the computation. Therefore, we define the following vectors at each cell boundary:

$$V = \begin{bmatrix} V^{\text{TEM}} \\ \vdots \\ V_{(\pm)}^{(m,i),e/h} \\ \vdots \\ V_{(\pm)}^{(M,I),e/h} \end{bmatrix} \quad \text{and} \quad I = \begin{bmatrix} I^{\text{TEM}} \\ \vdots \\ I_{(\pm)}^{(m,i),e/h} \\ \vdots \\ I_{(\pm)}^{(M,I),e/h} \end{bmatrix} \qquad (4)$$

One or more TEM modes exist if the background line is a multiconductor line (e.g., a coaxial cable), while they are absent if the background line is a single-conductor circular waveguide. In this paper, a coaxial cable is used as background line as shown in Figure 1, and a TEM mode is, therefore, present.

Therefore, the unit cell can be viewed as a multiport network: Each port corresponds to a mode at one of the periodic boundaries. We can define a multimodal generalized transmission matrix (or "T matrix") $\underline{T}$, relating the voltages and currents on one cell boundary to the voltages and currents on the other cell boundary [20]. The Floquet condition from Equation (2) can be stated in terms of this T matrix as

$$\underline{T}\cdot\begin{bmatrix} V \\ I \end{bmatrix} = e^{-jk_z p}\begin{bmatrix} V \\ I \end{bmatrix} \qquad (5)$$

The T matrix is a generalization of the well-known ABCD matrix, used for two-port networks. The ABCD matrix and the usual Bloch analyses treated in Reference [19] are recovered here if the fundamental TEM mode is the only retained mode in Equation (4). Next, we apply Equation (5) to the previous twisted lines to confirm the connection between irreducibility and cell coupling through higher-order modes. In all the following examples, the CST frequency-domain solver is at first used to calculate the scattering parameters in the frequency range 0–20 GHz. Then the scattering matrix is transformed into the T matrix as explained in Reference [20], and Equation (5) is solved. Note that a different definition of voltages and currents associated to each mode would change the numerical values of the scattering matrix and then would change the transimission matrix, but this would not modify the results of the eigenvalue problem in Equation (5).

Figure 3a depicts the dispersion curves obtained with Equation (5) for the 2-fold *reducible* twisted cell studied in Figure 2a. It also plots the dispersion diagram obtained with a CST ES for comparison. We observe that, in this case, a monomodal T matrix (1 mode) is already enough to match the results obtained by CST ES. The results for a T matrix with 3 modes ($m = 0, \pm 1$ and $i = 1$ in Equation (3)) also matches the monomodal T matrix and the CST ES results. This demonstrates that adding an extra set of modes does not vary the already converged results. Figure 3b shows the same results for the associated non-twisted line. Also, in this case, one single mode is sufficient to reach convergence. This confirms that in a reducible case a single mode is sufficient to accurately calculate the dispersion diagram of both the twisted and non-twisted lines.

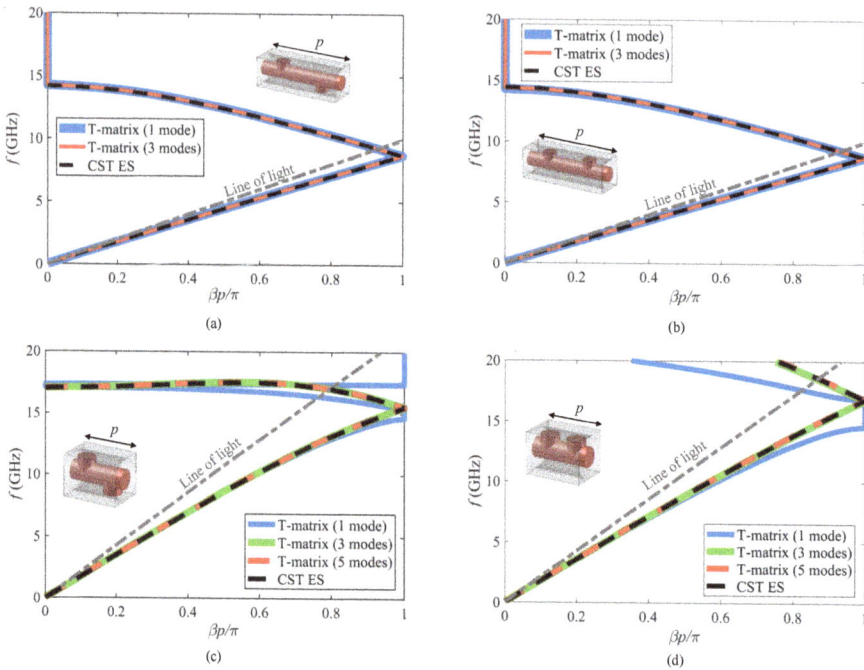

**Figure 3.** A dispersion diagram of 2-fold twist-symmetric structures in Figure 1 derived from the T-matrix method and CST ES applied to their unit cells: (**a**) Twisted $p = 15$ mm, (**b**) associated non-twisted $p = 15$ mm, (**c**) twisted $p = 7$ mm, and (**d**) associated non-twisted $p = 7$ mm.

Figure 3c depicts the dispersion diagram of the *irreducible* ($p = 7$ mm) 2-fold twisted unit cell studied in Figure 2b together with the dispersion diagram calculated with the CST ES. Here, a monomodal T-matrix method leads to the correct dispersion curve in the lower range of frequency. At higher frequencies, at least 3 modes are required. The results for 5 modes ($m = 0, \pm 1, \pm 2$ and $i = 1$) are also sketched in the figure to further emphasize the convergence. Figure 3d shows the same results for the associated non-twisted line. In this case, one mode only is not enough even at low frequencies. This confirms that the irreducibility of a twisted line is related to the presence of coupling through higher-order modes either in the twisted line or in its associated non-twisted line. These higher-order modes interact with each other differently in the presence or absence of twists and lead to different dispersive behaviors.

Figure 4a depicts the dispersion diagram for the 4-fold twisted cell *irreducible at high frequencies* studied in Figure 2c; Figure 4b shows the associated non-twisted line. Again, in the frequency ranges where the twisted line is irreducible, the presence of higher-order modes is necessary in at least one of the two lines (in this case the non-twisted line). This confirms that the difference between the lines arise from the presence of higher-order modes.

The same results are confirmed in Figure 4c,d (4-fold twisted cell *irreducible over the entire frequency range*): Higher-order modes are important in the non-twisted line over the entire range.

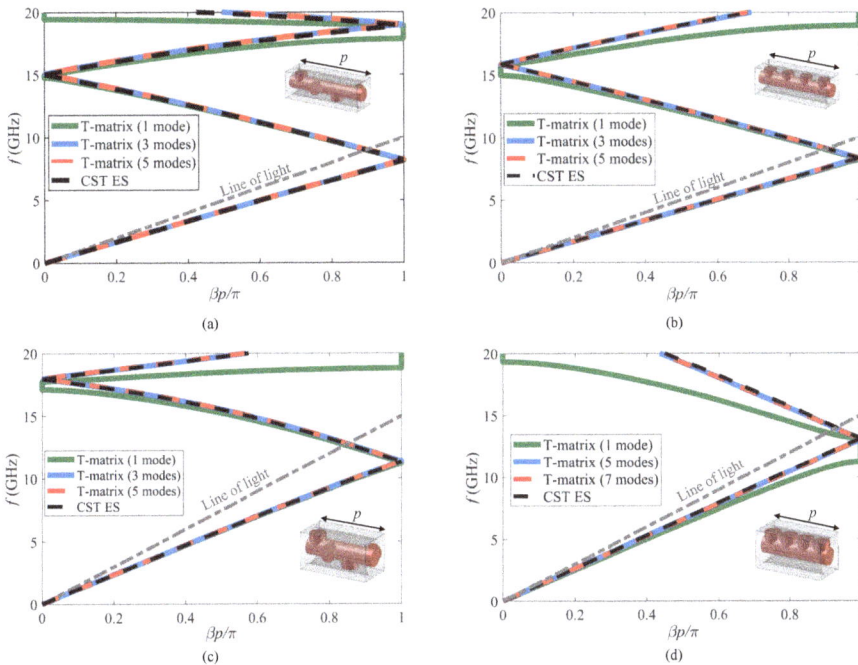

**Figure 4.** A dispersion diagram of 4-fold twist-symmetric structures in Figure 1 derived from the T-matrix method and CST ES applied to their unit cells: (**a**) Twisted $p = 15$ mm, (**b**) associated non-twisted $p = 15$ mm, (**c**) twisted $p = 10$ mm, and (**d**) associated non-twisted $p = 10$ mm.

These results confirm that, while reducible structures need a single fundamental mode, irreducible structures have a more complex modal interaction: Either they or their associated non-twisted line needs a multimodal T matrix. In this specific structure, the proximity of the pins is the key parameters that makes higher-order modes relevant. The more tightly coupled the pins are, the more relevant the higher-order modes become. For instance, a reducible twisted structure in Figure 1b could become an irreducible structure by decreasing the period and increasing the size of the pins since these changes will move the pins closer to each other.

We have shown that, in this context, modelling a unit cell with a multimodal T matrix is required for accurate dispersion results. This approach is also an effective alternative to retaining multiple unit cells as in Reference [21]. Finally, we stress that this method leads to accurate results also when eigenvalue tools of commercial software are currently not available, for example, in the case of open structures, where the attenuation constant $\alpha$ can be related to radiation (e.g., leaky waves).

## 3. Twist Symmetry Conditions on a Subunit Cell

In this section, we exploit the twist symmetry in order to reduce the computational domain of the periodic problem. In periodic lines, Bloch modes are eigensolutions of the translation operator $T_{p\hat{z}}$, as shown in Equation (2). By virtue of the twist symmetry, Bloch modes are also eigensolutions of the twist operator $S_{N,p\hat{z}}$ [4]:

$$S_{N,p\hat{z}}\left[E\left(\rho,\phi,z\right)\right] = \lambda E\left(\rho,\phi,z\right) \tag{6}$$

where $\lambda$ is the relevant eigenvalue. Note that the twist operator acts here on the observation point and does not rotate the $E$ field. Since $S_{N,p\hat{z}}^{N} = T_{p\hat{z}}$, from Equation (2), we can state that $\lambda^{N} = e^{-jk_{z}p}$ and that

$$\lambda = e^{-jk_{z}\frac{p}{N}} \tag{7}$$

where the different $N$th roots merely correspond to different Floquet harmonics, so that they do not appear in Equation (7).

The new eigenvalue problem in Equation (6) with $\lambda$ given in Equation (7) determines the Bloch modes. However, we are interested in formulating an eigenproblem for the translation operator $T_{\frac{p}{N}\hat{z}}$ rather than the twist operator. In fact, commercial software can easily compute a sub-cell transmission matrix, which describes the translation of fields and not the twist transformation. To overcome this problem, we first define the rotation operator:

$$R_{\frac{2\pi}{N}\hat{z}}\left[E\left(\rho,\phi,z\right)\right] = E\left(\rho,\phi+2\pi/N,z\right) \tag{8}$$

and we express the translation operator $T_{\frac{p}{N}\hat{z}}$ as a composition of the symmetry operator $S_{N,p\hat{z}}$ and the inverse rotation operator $R_{\frac{2\pi}{N}\hat{z}}^{-1} = R_{-\frac{2\pi}{N}\hat{z}}$:

$$T_{\frac{p}{N}\hat{z}}\left[E\left(\rho,\phi,z\right)\right] = R_{\frac{2\pi}{N}\hat{z}}^{-1}S_{N,p\hat{z}}E\left(\rho,\phi,z\right) = e^{-jk_{z}\frac{p}{N}}R_{\frac{2\pi}{N}\hat{z}}^{-1}E\left(\rho,\phi,z\right) \tag{9}$$

The T-matrix formulation requires the expression of the Bloch mode as a composition of the modes of the background structure. This is a convenient basis when dealing with rotations, since each background mode in Equation (3) satisfies a simple property:

$$R_{\frac{2\pi}{N}\hat{z}}^{-1}\left[\Psi_{(\pm)}^{(m,i)}\left(\rho,\phi\right)\right] = e^{\mp jm\frac{2\pi}{N}}\Psi_{(\pm)}^{(m,i)}\left(\rho,\phi\right) \tag{10}$$

(The $e/h$ polarization is not stated for simplicity). If the transmission matrix of a single *subunit* cell of an $N$-fold twisted structure associated to the translation operator $T_{\frac{p}{N}\hat{z}}$ is $\underline{T}_{1/N}$, the conditions in Equations (9) and (10) lead to the following eigenvalue problem:

$$\underline{T}_{1/N}\begin{bmatrix} V \\ I \end{bmatrix} = e^{-jk_{z}\frac{p}{N}}\begin{bmatrix} \underline{Q} & \underline{0} \\ \underline{0} & \underline{Q} \end{bmatrix}\cdot\begin{bmatrix} V \\ I \end{bmatrix} \tag{11}$$

where $V$ is the voltage and $I$ is the current vector defined in Equation (4). The matrix $\underline{Q}$ can be written as

$$\underline{Q} = \begin{bmatrix} 1 & 0 & \cdots & 0 \\ 0 & q_{(\pm)}^{(m,i)} & \ddots & \vdots \\ \vdots & \ddots & \ddots & 0 \\ 0 & \cdots & 0 & q_{(\pm)}^{(M,I)} \end{bmatrix} \tag{12}$$

where $q_{(\pm)}^{(m,i)} = e^{\mp jm\frac{2\pi}{N}}$ and $\underline{0}$ is the null square matrix.

Solving the generalized eigenvalue problem in Equation (11) gives the propagation constant $k_z$ starting from the simulation of a subunit cell of the twisted line. This formulation reduces the volume of the computational domain of the periodic problem by a factor $N$.

However, commercial electromagnetic software often calculates scattering parameters by means of waveguide modes with an angular dependence of the trigonometric forms $\cos(m\phi)$ or $\sin(m\phi)$ rather than the exponential form of $e^{\pm jm\phi}$. Therefore, a generalized eigenvalue problem based on these functions is useful for a practical implementation of the method. The radial components of the electric fields of these trigonometric modes are

$$\Psi_{\cos}^{(m,i),e/h}(\rho,\phi) = g_{(m,i)}^{e/h}(\rho)\cos(m\phi)$$

$$\Psi_{\sin}^{(m,i),e/h}(\rho,\phi) = g_{(m,i)}^{e/h}(\rho)\sin(m\phi) \tag{13}$$

where $g_{(m,i)}^{e/h}$ are suitable radial functions. Applying the inverse rotation operator to these electromagnetic fields gives the following (note again that the rotation is performed on the observation point, not on the vector field):

$$R_{\frac{2\pi}{N}\hat{z}}^{-1}\left[\Psi_{\cos}^{(m,i),e/h}(\rho,\phi)\right] = \left[\cos\left(\frac{2\pi m}{N}\right)\Psi_{\cos}^{(m,i),e/h}(\rho,\phi) + \sin\left(\frac{2\pi m}{N}\right)\Psi_{\sin}^{(m,i),e/h}(\rho,\phi)\right] \tag{14}$$

$$R_{\frac{2\pi}{N}\hat{z}}^{-1}\left[\Psi_{\sin}^{(m,i),e/h}(\rho,\phi)\right] = \left[-\sin\left(\frac{2\pi m}{N}\right)\Psi_{\cos}^{(m,i),e/h}(\rho,\phi) + \cos\left(\frac{2\pi m}{N}\right)\Psi_{\sin}^{(m,i),e/h}(\rho,\phi)\right] \tag{15}$$

It is interesting to note that the rotation (and then also the twist condition) is no more diagonal in the trigonometric basis: Rotating one mode gives a combination of two degenerate modes, with the only exception occurring for $m = 0$, where only the cos term exists. This means that we have to retain both the cos and the sin modes for each $m \neq 0$.

The generalized eigenvalue in Equation (11) changes to

$$\underline{T}_{1/N}\begin{bmatrix} V' \\ I' \end{bmatrix} = e^{-jk_z\frac{p}{N}}\begin{bmatrix} Q' & 0 \\ 0 & Q' \end{bmatrix} \cdot \begin{bmatrix} V' \\ I' \end{bmatrix} \tag{16}$$

where the primed voltage and current vectors $V'$ and $I'$ are defined based on the trigonometric background modes:

$$V' = \begin{bmatrix} V^{\text{TEM}} \\ \vdots \\ V_{\cos}^{(m,i),e/h} \\ V_{\sin}^{(m,i),e/h} \\ \vdots \\ V_{\cos}^{(M,I),e/h} \\ V_{\sin}^{(M,I),e/h} \end{bmatrix} \quad \text{and} \quad I' = \begin{bmatrix} I^{\text{TEM}} \\ \vdots \\ I_{\cos}^{(m,i),e/h} \\ I_{\sin}^{(m,i),e/h} \\ \vdots \\ I_{\cos}^{(M,I),e/h} \\ I_{\sin}^{(M,I),e/h} \end{bmatrix} \tag{17}$$

The definition of the primed matrix $Q'$ is also given here:

$$
\underline{Q}' = \begin{bmatrix}
1 & 0 & \cdots & & \cdots & & \cdots & & \cdots & & 0 \\
0 & \ddots & & \ddots & & \ddots & & \cdots & & \cdots & \vdots \\
\vdots & \ddots & \cos\left(\frac{2\pi m}{N}\right) & \sin\left(\frac{2\pi m}{N}\right) & \cdots & & \cdots & & & \vdots \\
\vdots & \ddots & -\sin\left(\frac{2\pi m}{N}\right) & \cos\left(\frac{2\pi m}{N}\right) & \cdots & & \cdots & & & \vdots \\
\vdots & \ddots & \ddots & & \ddots & & \ddots & 0 & & 0 \\
\vdots & \ddots & \ddots & & & \ddots & & 0 & \cos\left(\frac{2\pi M}{N}\right) & \sin\left(\frac{2\pi M}{N}\right) \\
0 & \cdots & & \cdots & & \cdots & & 0 & -\sin\left(\frac{2\pi M}{N}\right) & \cos\left(\frac{2\pi M}{N}\right)
\end{bmatrix} \tag{18}
$$

Figure 5a plots the normalized phase constant of the 4-fold twist-symmetric structure shown in Figure 1b ($p = 15$ mm) derived from Equation (16) and compares it to the results given by CST ES. It is observed that a single mode does not provide the correct dispersion diagram at high frequencies, whereas considering 3 modes (TEM mode and the first two degenerate higher-order modes) leads to the accurate diagram, which matches the CST ES results over the entire frequency band shown. To further emphasize the advantages of the T-matrix method, the normalized attenuation constant of the structure is also given in Figure 5b where it is easy to notice that the stopband predicted with a single mode is not correct while using 3 modes correctly predicts the frequency range of first stopband. These results could not be compared to CST ES since the commercial software does not provide such information. The results with inclusion of 5 modes are also plotted in these two figures, and they match the results of the 3-mode T-matrix method. It is important to note that even though we have used the pin-loaded transmission lines in Figure 1 to verify our method, the eigenvalue problems in Equations (11) and (16) are general and they can be applied to any structure with a twist symmetry.

Furthermore, the method is also valid for a larger class of structure, characterized by the invariance under the twist transformation of Equation (1) with non-integer $N$. In this last case, the structure is no more periodic and, therefore, cannot be studied with available commercial software. The present approach still holds and, to the best of the authors' knowledge, is the only formulation available for a rigorous solution of the problem.

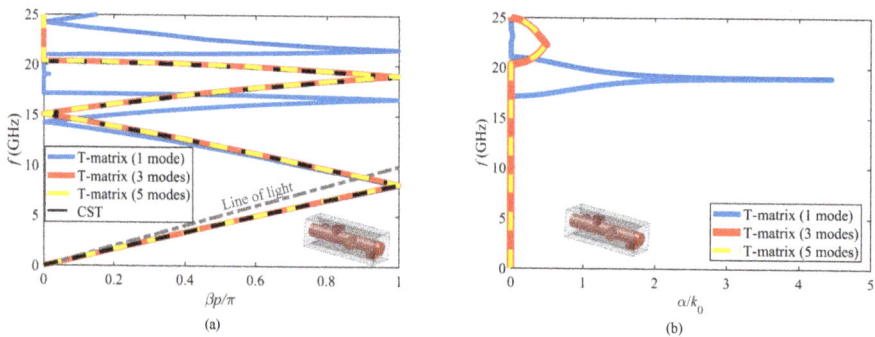

**Figure 5.** Dispersion diagrams derived with the T-matrix method on the subunit cell of the structure in Figure 1b ($p = 15$ mm): (**a**) The normalized phase constant $\beta p / \pi$ vs. frequency and (**b**) the normalized attenuation constant $\alpha / k_0$ vs. frequency ($k_0$ being the free-space wavenumber).

Finally, it is interesting to compare the solving time of the T-matrix method applied to a sub-cell and the CST ES in an entire unit cell. To do so, we consider the more complex case of the 4-fold twist symmetry where 3 modes were needed in the T-matrix method. To make this comparison, we choose 76 points on the first passband and solve the problem for this frequency range. For the CST ES, an adaptive mesh refinement with a maximum of 1% relative error in frequency is defined using tetrahedral meshes. As a result, the cell is meshed with 19,850 tetrahedrons. For the CST frequency-domain solver, a relative error of 1% for scattering parameters is defined in the adaptive mesh refinement routine with tetrahedral meshes, resulting in a total of 14,310 tetrahedrons for the entire cell and 4016 tetrahedrons for the sub-cell. We have used a computer with 128 GB of RAM and an Intel(R) Xeon(R) CPU with 6 cores and a base frequency of 3.60 GHz for its CPU cores.

The computation time for these methods are included in Table 1. A quick comparison for the case of a single frequency point shows that even though the T-matrix method with a unit cell has a higher computation time compared to CST ES, the T-matrix method with the subunit cell is slightly faster than CST ES. It should be noted that a higher computation time of the unit cell formulation may be justified since the T matrix generally produces extra information on the attenuation constant of the Bloch waves and can be applied also to open structures. As we consider a large frequency range, the computation time difference changes dramatically. The T-matrix method has a shorter computation time for both the unit cell and subunit cell formulations (88 and 21 s, respectively) compared to CST ES (780 s). This occurs because the computation of the scattering parameters is accelerated significantly by the broadband sweep techniques. The rest of the computation (solving the T-matrix eigenvalue problem) is executed very fast. For instance, this part was executed in less than 2 s for 76 frequency points and 3 modes for both the unit cell and the subunit cell formulations. This is a significant reduction in computation time compared to the CST ES.

**Table 1.** The computational time for solving the periodic structure in Figure 1b.

| 1 frequency point | T matrix (subunit Cell) | T matrix (unit cell) | CST (eigensolver) |
|---|---|---|---|
| Time (s) | 12 | 19 | 13 |
| 76 frequency points | T matrix (subunit cell) | T matrix (unit cell) | CST (eigensolver) |
| Time (s) | 21 | 88 | 780 |

## 4. Conclusions

In this paper, we have discussed a transmission-matrix method for periodic structures with a higher symmetry. We have enhanced its accuracy by including higher-order modes and have demonstrated through a set of examples how the inclusion of these higher modes increases the accuracy of the method. Furthermore, we have used higher-order modes to explain how some twisted structures with symmetry can be equivalent to non-twisted structures with a reduced period. We also applied the twist operator to a transmission matrix method for the first time. This resulted in a new formulation that applies to a subregion of a unit cell for structures with a twist symmetry with a lower computation cost.

**Author Contributions:** Conceptualization, G.V.; methodology, M.B and G.V.; software, M.B.; validation, M.B.; formal analysis, M.B. and G.V.; writing—original draft preparation, M.B.; writing—review and editing, M.B. and G.V.; visualization, M.B.; supervision, G.V.; project administration, G.V.; funding acquisition, G.V.

**Funding:** This work was supported by the French governement under the ANR grant HOLeYMETA ANR JCJC 2016 ANR-16-CE24-0030 and by Sorbonne Universités under the Emergence 2016 grant MetaSym.

**Acknowledgments:** The authors wish to thank Francisco Mesa and Raúl Rodríguez-Berral from Universidad de Sevilla for useful discussions on the subject.

**Conflicts of Interest:** The authors declare no conflict of interest.

# References

1. Crepeau, P.J.; McIsaac, P.R. Consequences of symmetry in periodic structures. *Proc. IEEE* **1964**, *52*, 33–43. [CrossRef]
2. Mittra, R.; Laxpati, S. Propagation in a wave guide with glide reflection symmetry. *Can. J. Phys.* **1965**, *43*, 353–372. [CrossRef]
3. Kieburtz, R.; Impagliazzo, J. Multimode propagation on radiating traveling-wave structures with glide-symmetric excitation. *IEEE Trans. Antennas Propag.* **1970**, *18*, 3–7. [CrossRef]
4. Hessel, A.; Chen, M.H.; Li, R.C.; Oliner, A.A. Propagation in periodically loaded waveguides with higher symmetries. *Proc. IEEE* **1973**, *61*, 183–195. [CrossRef]
5. Chen, Q.; Ghasemifard, F.; Valerio, G.; Quevedo-Teruel, O. Modeling and Dispersion Analysis of Coaxial Lines with Higher Symmetries. *IEEE Trans. Microw. Theory Tech.* **2018**, *66*, 4338–4345. [CrossRef]
6. Quevedo-Teruel, O.; Dahlberg, O.; Valerio, G. Propagation in waveguides with transversal twist-symmetric holey metallic plates. *IEEE Microw. Wirel. Compon. Lett.* **2018**, *28*, 858–860. [CrossRef]
7. Dahlberg, O.; Mitchell-Thomas, R.C.; Quevedo-Teruel, O. Reducing the Dispersion of Periodic Structures with Twist and Polar Glide Symmetries. *Sci. Rep.* **2017**, *7*. [CrossRef] [PubMed]
8. Ghasemifard, F.; Norgren, M.; Quevedo-Teruel, O. Twist and polar glide symmetries: an additional degree of freedom to control the propagation characteristics of periodic structures. *Sci. Rep.* **2018**, *8*, 11266. [CrossRef] [PubMed]
9. Quesada, R.; Martín-Cano, D.; García-Vidal, F.J.; Bravo-Abad, J. Deep-subwavelength negative-index waveguiding enabled by coupled conformal surface plasmons. *Opt. Lett.* **2014**, *39*, 2990–2993. [CrossRef] [PubMed]
10. Camacho, M.; Mitchell-Thomas, R.C.; Hibbins, A.P.; Sambles, J.R.; Quevedo-Teruel, O. Designer surface plasmon dispersion on a one-dimensional periodic slot metasurface with glide symmetry. *Opt. Lett.* **2017**, *42*, 3375–3378. [CrossRef] [PubMed]
11. Camacho, M.; Mitchell-Thomas, R.C.; Hibbins, A.P.; Sambles, J.R.; Quevedo-Teruel, O. Mimicking glide symmetry dispersion with coupled slot metasurfaces. *Appl. Phys. Lett.* **2017**, *111*, 121603. [CrossRef]
12. Palomares-Caballero, A.; Padilla, P.; Alex-Amor, A.; Valenzuela-Valdés, J.; Quevedo-Teruel, O. Twist and Glide Symmetries for Helix Antenna Design and Miniaturization. *Symmetry* **2019**, *11*, 349. [CrossRef]
13. Quevedo-Teruel, O.; Ebrahimpouri, M.; Kehn, M.N.M. Ultrawideband Metasurface Lenses Based on Off-Shifted Opposite Layers. *IEEE Antennas Wirel. Propag. Lett.* **2016**, *15*, 484–487 [CrossRef]
14. Ebrahimpouri, M.; Rajo-Iglesias, E.; Sipus, Z.; Quevedo-Teruel, O. Cost-Effective Gap Waveguide Technology Based on Glide-Symmetric Holey EBG Structures. *IEEE Trans. Microw. Theory Tech.* **2018**, *66*, 927–934. [CrossRef]
15. Rajo-Iglesias, E.; Ebrahimpouri, M.; Quevedo-Teruel, O. Wideband Phase Shifter in Groove Gap Waveguide Technology Implemented With Glide-Symmetric Holey EBG. *IEEE Microw. Wirel. Compon. Lett.* **2018**, *28*, 476–478. [CrossRef]
16. Wu, J.J.; Wu, C.J.; Hou, D.J.; Liu, K.; Yang, T.J. Propagation of Low-Frequency Spoof Surface Plasmon Polaritons in a Bilateral Cross-Metal Diaphragm Channel Waveguide in the Absence of Bandgap. *IEEE Photonics J.* **2015**, *7*, 1–8. [CrossRef]
17. Ebrahimpouri, M.; Brazalez, A.A.; Manholm, L.; Quevedo-Teruel, O. Using Glide-Symmetric Holes to Reduce Leakage Between Waveguide Flanges. *IEEE Microw. Wirel. Compon. Lett.* **2018**, *28*, 473–475. [CrossRef]
18. CST Microwave Studio. Version: 2019. Available online: http://www.cst.com/ (accessed on 15 March 2019).
19. Pozar, D.M. *Microwave Engineering*, 4th ed.; John Wiley & Sons: Hoboken, NJ, USA, 2012.
20. Collin, R. *Field Theory of Guided Waves*; Wiley-IEEE Press: Hoboken, NJ, USA, 1991.
21. Valerio, G.; Paulotto, S.; Baccarelli, P.; Burghignoli, P.; Galli, A. Accurate Bloch Analysis of 1-D Periodic Lines Through the Simulation of Truncated Structures. *IEEE Trans. Antennas Propag.* **2011**, *59*, 2188–2195. [CrossRef]

MDPI

St. Alban-Anlage 66

4052 Basel

Switzerland

Tel. +41 61 683 77 34

Fax +41 61 302 89 18

www.mdpi.com

*Symmetry* Editorial Office

E-mail: symmetry@mdpi.com

www.mdpi.com/journal/symmetry

www.ingramcontent.com/pod-product-compliance
Lightning Source LLC
Chambersburg PA
CBHW051916210326
41597CB00033B/6161